AF839605

Swarm Intelligence

Theory and Applications in Fog Computing, Beyond 5G Networks, and Information Security

Editors

V. Dinesh Reddy

Md. Muzakkir Hussain

Priyanka Singh

Computer Science and Engineering, SRM University
Mangalagiri-Mandal, Amaravati, India

CRC Press
Taylor & Francis Group
Boca Raton London New York

CRC Press is an imprint of the
Taylor & Francis Group, an **informa** business
A SCIENCE PUBLISHERS BOOK

First edition published 2025
by CRC Press
2385 NW Executive Center Drive, Suite 320, Boca Raton FL 33431

and by CRC Press
4 Park Square, Milton Park, Abingdon, Oxon, OX14 4RN

© 2025 V. Dinesh Reddy, Md. Muzakkir Hussain and Priyanka Singh

CRC Press is an imprint of Taylor & Francis Group, LLC

Reasonable efforts have been made to publish reliable data and information, but the author and publisher cannot assume responsibility for the validity of all materials or the consequences of their use. The authors and publishers have attempted to trace the copyright holders of all material reproduced in this publication and apologize to copyright holders if permission to publish in this form has not been obtained. If any copyright material has not been acknowledged please write and let us know so we may rectify in any future reprint.

Except as permitted under U.S. Copyright Law, no part of this book may be reprinted, reproduced, transmitted, or utilized in any form by any electronic, mechanical, or other means, now known or hereafter invented, including photocopying, microfilming, and recording, or in any information storage or retrieval system, without written permission from the publishers.

For permission to photocopy or use material electronically from this work, access www.copyright.com or contact the Copyright Clearance Center, Inc. (CCC), 222 Rosewood Drive, Danvers, MA 01923, 978-750-8400. For works that are not available on CCC please contact mpkbookspermissions@tandf.co.uk

Trademark notice: Product or corporate names may be trademarks or registered trademarks and are used only for identification and explanation without intent to infringe.

Library of Congress Cataloging-in-Publication Data (applied for)

ISBN: 978-1-032-67697-5 (hbk)
ISBN: 978-1-032-69320-0 (pbk)
ISBN: 978-1-032-69321-7 (ebk)

DOI: 10.1201/9781032693217

Typeset in Times New Roman
by Prime Publishing Services

To my family
and to all the contributors.

Preface

In the changing world of information technology the combination of Swarm Intelligence and Fog Computing has emerged as a source of innovation. This book, titled "Swarm Intelligence: Theory and Applications in Fog Computing, Beyond 5G Networks, and Information Security" provides an exploration of topics that contribute to our understanding, development and ethical use of these advanced technologies. The opening chapter sets the stage by diving, into the concept of "Fog Intelligence" offering an analysis of how swarm algorithms can be utilized to achieve a high level of performance in edge computing. Fog computing, known for its architecture brings computation closer to the source of data. By integrating swarm intelligence into this framework we have the potential to revolutionize the efficiency and adaptability of edge computing systems. Moving forward the next chapter examines how swarm intelligence can be applied within the context of Internet of Things (IoT) and Edge Computing. As IoT continues to expand there is an increasing need for adaptable systems. The combination of swarm intelligence with edge computing offers solutions to enhance scalability, reliability and autonomy among interconnected devices. The third chapter introduces a metaheuristic based approach for optimizing the deployment of applications, on fog computing platforms. It is essential to conduct this investigation to ensure utilization of the fog environment maximizing resource allocation and enhancing the performance of the system. The intricate interplay of algorithms holds the potential to navigate the intricacies of deploying applications providing a roadmap, for the development of IoT systems based on fog computing.

The fourth chapter takes us into the world of wireless powered UAV IoT networks, where swarm intelligence is utilized for dynamic task allocation. This innovative approach not only helps optimize energy usage but also improves the efficiency of these networks demonstrating the transformative potential of swarm algorithms, in real life applications. The fifth chapter focuses on disaster management. It conducts a survey on the utilization of UAV swarms in search and rescue operations. By incorporating swarm intelligence into vehicles their capabilities are enhanced, making them invaluable tools in critical situations. This chapter showcases advancements but it also highlights the humanitarian aspect of swarm intelligence in responding to disasters.

Exploring the intersection between machine learning, deep learning and swarm algorithms the sixth chapter examines how these fields work together. As artificial intelligence continues to advance combining learning methods with swarm intelligence introduces approaches to problem solving, pattern recognition and decision making. Moving beyond networks the seventh chapter delves into the complexities surrounding privacy in swarm embedded systems. Specifically focusing on 5G networks this article explores topics such, as data anonymization, secure data transportation and user consent. The discussion goes beyond considerations by emphasizing dimensions and stressing the importance of responsible and secure implementation of swarm intelligence in eighth chapter. In this chapter the issue of privacy remains a concern. The chapter delves into safeguarding user privacy within distributed computing networks at the edge of fog computing. With the growing trend of data processing it offers insights, into maintaining privacy without making any compromises to the advantages of fog computing. The ninth chapter focuses on the aspect of security presenting an authentication scheme specifically tailored for fog centric IoT device systems. In todays interconnected world, where data integrity is of importance this chapter offers a strong solution to strengthen IoT systems against potential threats. The book concludes with the chapter, which delves into the challenges associated with swarm intelligence, in fog computing. It provides frameworks to deploy these technologies and encourages an conscientious approach to their development and implementation. Ultimately “Swarm Intelligence and Fog Computing” is more than a collection of topics; it takes readers on a cohesive journey through innovation, technology and ethics. Each chapter contributes a perspective to the overarching narrative of how swarm intelligence and fog computing are converging. Whether you’re a researcher, or in industry, or someone interested in understanding the future of technology this book serves as a guide to navigate this intersection of innovation.

Acknowledgment

The guest editors would like to thank all authors for their willingness to work on these complex book projects.We thank the authors for their interesting proposals of the book chapters, their time, efforts and their research results, which makes this volume an interesting complete monograph of the latest research advances and technology development in the areas of fog computing and swarm intelligence. Our special thanks go to the editorial and managerial team from Taylor & Francis for their continuous support and excellent cooperative collaboration to produce this valuable scientific work. Finally, we would like to send our warmest gratitude message to our friends and families for their patience, love, and support in the preparation of this volume. We strongly believe that this book ought to serve as a reference for students, researchers, and industry practitioners interested or currently working in swarm intelligence, 5g networks, IoT and the fog computing domains.

Contents

List of Figures

List of Tables

Chapter 1

Resource Management in Fog Computing:

Overview and Mathematical Foundation

Surayya, A.,[1] *Md. Muzakkir Hussain,*[1,*] *V. Dinesh Reddy,*[1] *Ahsan Halimi*[2] and *Firoj Gazi*[1]

Fog computing is a distributed computing paradigm that extends the capabilities of cloud computing to the edge of the network, closer to the data source or user. Resource management in fog computing is a complex task due to the heterogeneity of devices, dynamic workloads, limited resources, energy efficiency, task offloading, load balancing, quality of service (QoS) management, security, and privacy concerns. It plays a crucial role in optimizing the performance and efficiency of fog computing systems. The chapter delves into the challenges posed by the diverse nature of devices, dynamic workloads, and distributed architecture, emphasizing the need for adaptive resource allocation strategies. It provides a systematic and mathematical approach to resource management, including the formulation of optimization problems such as the Knapsack Problem, Traveling Salesman Problem, Transportation Problem, Vehicular Routing Problem, and N-Queens Problem. Furthermore, it underscores the significance of load balancing,

[1] Department of Computer Science and Engineering, SRM University, AP, India.
[2] Department of Electronics and Communication Engineering, VR Siddhartha Engineering College, Vijayawada, India.
* Corresponding author: muzakkirhussain.m@srmap.edu.in

task offloading, and resource provisioning as adaptive strategies to dynamically allocate resources, ensuring optimal utilization without causing underutilization. It offers valuable insights into the complexities of managing resources in fog computing and provides a holistic view of the challenges, strategies, and mathematical formulations involved in resource management across various contexts.

1.1 Introduction

Resource management in fog computing is a pivotal aspect that plays a central role in optimizing the performance, reliability, and scalability of this innovative paradigm [153]. Fog computing, an extension of cloud computing, strategically positions computational resources at the edge of the network, bringing processing capabilities closer to data sources and enabling real-time, low-latency processing [167]. This distributed architecture requires dynamic allocation, optimization, and monitoring of resources, considering the diverse and dynamic nature of edge devices. One of the fundamental challenges in resource management within fog computing arises from the heterogeneity of devices comprising fog environments [44]. With varying capabilities and characteristics, managing resources efficiently across different types of devices becomes a complex task. This diversity encompasses devices ranging from sensors and actuators to more powerful edge servers, demanding adaptive resource allocation strategies. Technological solutions, including machine learning and artificial intelligence, play a significant role in predictive resource management. These algorithms enable fog nodes to anticipate future workloads and adapt proactively, contributing to the overall efficiency of resource utilization. Containerisation and virtualization technologies, such as Docker and Kubernetes, facilitate efficient resource allocation by encapsulating applications into containers, ensuring consistency and portability across diverse edge devices.

The dynamic nature of workloads in fog computing introduces another layer of complexity. Workloads can be highly dynamic due to factors such as varying user demands, the mobility of devices, and changing network conditions. Effective resource management must, therefore, employ adaptive strategies to allocate resources dynamically, ensuring optimal utilization without causing bottlenecks or under utilization [86]. The constraint of limited resources on edge devices further complicates resource management. Many edge devices, including IoT sensors and actuators, operate with constrained resources in terms of processing power, memory, and energy [253]. Striking a balance between utilizing these resources effectively and preventing overload or under utilization becomes a critical consideration in fog computing environments. Energy efficiency emerges as a paramount concern in fog computing, particularly as numerous edge devices rely on battery power. Optimizing resource usage not only prolongs device lifespan but also aligns with the growing emphasis on environmentally sustainable computing practices. Therefore, resource management strategies must incorporate energy-efficient algorithms to address the unique challenges posed by the limited energy capacity of edge devices [111].

Several key aspects constitute the essence of resource management in fog computing. Task offloading stands out as a critical element, involving decisions on where to execute specific tasks – at the edge, in the cloud, or a hybrid combination [157]. Intelligent offloading decisions leverage optimization algorithms that consider factors such as computation and communication costs, contributing to overall system efficiency [258]. Load balancing is another key aspect, aiming to distribute tasks and workloads evenly across fog nodes. Dynamic load balancing algorithms adapt to changes in workload distribution, ensuring optimal resource utilization and preventing congestion in specific nodes [5]. Quality of Service (QoS) management is paramount in fog computing, where resource management strategies must consider QoS requirements such as latency, reliability, and bandwidth. These considerations are crucial for meeting the diverse expectations of applications ranging from real-time IoT applications to augmented reality. To address these requirements effectively, resource allocation and task scheduling decisions must be optimized to ensure that the necessary QoS levels are maintained throughout the fog computing environment [75]. Addressing security and privacy concerns is an integral part of resource management in fog computing. Implementing secure data storage, communication, and access control mechanisms safeguards sensitive information at the edge, ensuring data integrity and user privacy. By integrating security measures into resource management strategies, fog computing environments can mitigate potential risks associated with unauthorized access, data breaches, and malicious attacks, thereby fostering trust and confidence among users and stakeholders.

As fog computing continues to evolve, future directions in resource management involve advancements in adaptive algorithms, the integration of 5G networks for enhanced communication capabilities, and the development of standardized frameworks for interoperability. Overcoming challenges and embracing innovative technologies will shape the landscape of resource management in fog computing, solidifying its position as a transformative paradigm in the realm of edge computing [205]. Table 1.1 presents a comparison between edge computing (EC), cloud computing (CC), fog computing (FC), and vehicular fog computing (VFC) across various dimensions, including latency, resource location, scalability, mobility support, and security. Each computing paradigm is evaluated based on its performance and capabilities in these key areas, providing insights into their suitability for different applications and scenarios. In this chapter, we investigate resource management in vehicular fog computing, focusing on essential contributions that impact our understanding and optimization of resource management in dynamic situations. The key contributions in this chapter can be summarized as follows:

- Resource categorization: We have categorized the resources in fog computing into various types and provided a mathematical representation of each.
- Resource profiling: This section covers the following concepts: task, decision variable, objective functions and constraints.
- Resource management sub-problems: It involves eight main fields including task offloading, application placement, resource scheduling, load balancing, resource allocation, resource provisioning, service work flow, resource orchestration, and resource pooling.
- Optimization problems: We focused on the different types of optimization problems include linear programming problems, nonlinear programming problems, quadratic problems, integer programming, mixed-integer programming, and dynamic programming.
- Solving resource management problems using reduction to standard optimization problems: We explored how resource management problems can be addressed by comparing them to five essential benchmark problems.
- Case study: This study explores the use of vehicular fog computing for real-time object detection in automatic vehicles (AVs), specifically focusing on a critical scenario involving a toddler crossing the road.

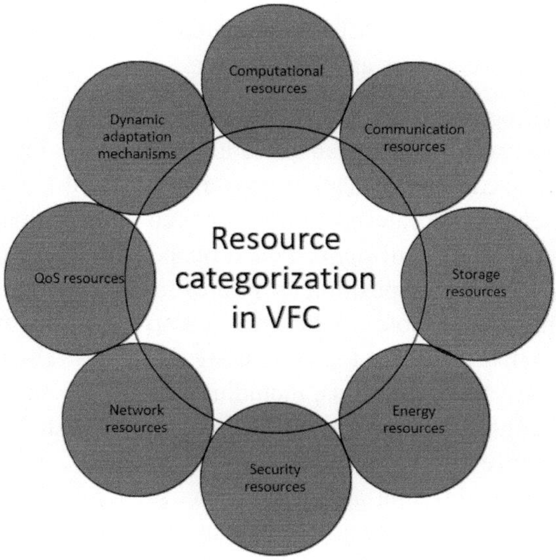

Figure 1.1: Resource categorization in fog computing.

Table 1.1: Comparison of computing models that are similar to FC.

Attribute	Cloud computing (CC)	Mobile Cloud computing (MCC)	Edge computing (EC)	Mobile Edge computing (MEC)	Fog computing (FC)	Vehicular Fog computing (VFC)
General Use Case	Scalable data storage, virtualized apps, urban traffic management	Social Internet of Vehicles (SIoV)	Local video surveillance, video caching, traffic control	Disaster response networks	Smart City	Connected Vehicles, drone Delivery, video Surveillance, vehicle safety
Operators	CSPs	Mobile users and cloud service providers	Network infrastructure providers or local IoT	Radio access network (RAN) service providers	Users as well as cloud service providers	Vehicles, CSPs
Hardware	CDCs	Mobile Devices and CDCs	Edge devices	With computing capability	Micro-data centers, C-RAN	Servers, routers, switches, access points, F-RAN
Architecture	Centralized/ hierarchical	Central cloud with distributed mobile devices	Localized/distributed	Localized/ Hierarchical	Decentralized/Hierarchical	Decentralized/Hierarchical
Availability	High	High	Average	Average	High	High
Latency	Relatively high	Relatively High	Low	Low	Low	Very Low
Security	Must be provided along cloud-to-things continuum	Must be provided along cloud to things continuum and on mobile devices	Must be provided on edge devices	Must be provided on edge network equipment (RAN, AP)	Must be provided on participant nodes	Must be provided on edge network equipment (RAN, AP)
Power Consumption	Low on mobile devices	Low	High	Low	Low	Low
Main Standardization Entity	CSA, DMFT, NIST, OCC, GICTF	NIST	ETSI, 3GPP, ITU-T	OpenFog, IEEE	VSC, OpenFog	VSC, OpenFog

1.2 Resource Categorization in Fog Computing

In the context of fog computing and resource management, resources typically refer to the various elements within the fog computing infrastructure that are allocated, utilized, and optimized to support the execution of applications and services [105]. The different types of resources are depicted in Figure 1.1. These resources can be both hardware and software components essential for the functioning of the fog computing environment. The following are some key types of resources in resource management:

- **Computational Resources:** The main computational resources are CPU and memory usage.The CPU processing power represents the computing capability required to execute applications and process data [37]. Memory (RAM or ROM) storage is used for temporarily holding data and program instructions during processing. Mathematically, the computational resources can be represented as a tuple: $< \rho, \mu >$

  where ρ represents the processing power and μ represents the available memory of the fog node. In the case of VFC, the fog node may be either dedicated hardware called a static fog node (SFN) or moving nodes called a dynamic fog node (DFN).

- **Communication Resources:** The key communication resources are network bandwidth and latency. Bandwidth is a measure of the capacity of the communication channels to transmit data between edge devices, fog nodes, devices, and the cloud. Network latency,defined as the time delay in data transmission, is crucial for real-time applications. The bandwidth and latency between fog nodes are represented by tuple: $\langle \lambda, \beta \rangle$, which also serves as a representation of the communication resources. In the case of VFC, the band width is associated with the V2X communication links. The key technologies are 802.11p DSRC (75 MHz), LTE (1.4 MHz, 3 MHz, 5 MHz, 10 MHz, 15 MHz, and 20 MHz), C-V2X, 5G NR V2X (5 MHz, 10 MHz, 20 MHz, 40 MHz, 50 MHz, and 100 MHz), etc.

- **Storage Resources:** It offers two main options for managing data: local storage and distributed storage. With local storage, data can be stored directly on fog nodes, allowing applications to access it quickly when needed. On the other hand, distributed storage spreads data across several fog nodes, which helps to improve the system's ability to withstand faults and ensure data remains accessible. These storage strategies play a crucial role in optimizing performance and reliability within fog computing networks. The representation of the storage resources is: $< \alpha, \delta >$. The fog node's local storage capacity is represented by α .The distributed storage capacity among fog nodes is represented by δ.

- **Energy Resources:** The power supply and battery life are the two components of energy resources. Power supply is important for providing energy to fog nodes and connected devices. It ensures that these devices have the power they need to function properly. Battery life is specifically crucial for mobile devices and vehicles in vehicular fog computing situations. It determines how long these devices can operate without needing to be recharged or connected to a power source. Energy resources are denoted by: $< P, B >$ where P specifies the availability of power supply for the fog node and B denotes the amount of battery life left in mobile devices or cars. Energy consumption due to communication and computation contains transmission energy, computation energy, power consumption, task size, and upload data rate.
- **Security Resources:** Two essential resources are cryptography tools and authentication and authorization systems. Cryptography tools serve as foundational resources for implementing robust security measures, facilitating secure communication channels, and using data encryption to safeguard sensitive information against unauthorized access or tampering. Authentication and authorization systems play a pivotal role in managing access control and permissions, ensuring that only authorized users or devices can access specific resources or functionalities within the fog computing environment. Security resources are denoted as: $< C, A >$, where C denotes the availability of cryptography tools on the fog node. The authentication and authorization are represented by A.
- **Network Resources:** The primary network resources are network protocols and routing and switching capabilities. Routing and switching capabilities help direct information within the network efficiently. Network protocols ensure that communication between fog nodes and devices follows a standardized approach. It can be represented as: $< \Theta, \psi >$, where Θ reflects the fog node routing and switching capabilities, and ψ denotes the network protocols supported by the fog node.
- **QoS (Quality of Service) Resources:** It includes prioritization mechanisms and traffic shaping. Priority mechanisms for identifying critical applications or services and traffic shaping for controlling data flow to meet specific QoS requirements. QoS resources can be represented as: $< \phi, \zeta >$. Where φ represents the prioritization mechanism and ζ represents the traffic shaping mechanism on the fog node.
- **Dynamic Adaptation Mechanisms:** Dynamic adaptation mechanisms involve two key components: dynamic scaling and resource orchestration. Dynamic scaling is the process of adjusting resource allocation in response to changes in workload or demand. Resource orchestration, on the other hand, focuses on the intelligent allocation and de-allocation of

resources to optimize performance. Dynamic adaptation mechanisms are represented as: $< D,R >$, where D represents the dynamic scaling capability between fog nodes and R represents the resource orchestration capability.

1.3 Resource Profiling

Resource profiling in edge computing involves the systematic analysis and characterization of the computational resources available in edge devices [240]. This process aims to understand the capabilities, limitations, and usage patterns of individual devices within an edge computing environment. Resource profiling plays a crucial role in optimizing the allocation and utilization of resources, enhancing overall system performance, and supporting efficient application deployment. Here are key aspects and considerations related to resource profiling in edge computing:

- **Device Capabilities and Constraints:** Resource profiling begins by assessing the hardware and software capabilities of edge devices. This includes examining the processing power, memory, storage, and communication interfaces of each device. Understanding the constraints and limitations, such as energy constraints in battery-powered devices, is essential for effective resource management.

- **Network Connectivity:** Profiling network connectivity is crucial in edge computing, as it influences communication latency and bandwidth. Analyzing the network capabilities helps in determining the communication patterns between edge devices and the central infrastructure. This information is vital for applications relying on real-time data processing and low-latency communication.

- **Workload Patterns:** Profiling involves studying the historical and dynamic workload patterns on edge devices. This includes analyzing the types of tasks or applications typically executed on each device, the duration of tasks, and the frequency of task arrivals. Workload profiling aids in predicting resource demands, facilitating proactive resource allocation and load balancing.

- **Energy Consumption:** Energy efficiency is a significant concern in edge computing, particularly for devices with limited power sources. Profiling energy consumption patterns helps in identifying power-hungry tasks or applications. This information is valuable for optimizing task placement and scheduling to enhance overall energy efficiency.

- **Security and Privacy Considerations:** Profiling should also address security and privacy aspects. Understanding the security features and vulnerabilities of edge devices is crucial for deploying secure applications. Additionally, profiling may involve assessing the privacy implications of processing sensitive data on specific devices.

- **Application Requirements:** Profiling extends to understanding the resource requirements of specific applications. Different applications may have varying demands for computational power, memory, and storage. Profiling these requirements aids in matching applications with suitable edge devices, ensuring optimal performance.
- **Dynamic Adaptability:** Edge environments are dynamic, and resource profiling should support adaptability to changes in the system. This includes accommodating variations in device availability, network conditions, and workload intensity. Dynamic profiling enables real-time adjustments to resource allocation strategies.
- **Tooling and Automation:** Employing tools and automation for resource profiling streamlines the data collection and analysis process. Various profiling tools can collect information about device characteristics, network performance, and application behavior. Automation ensures the continuous monitoring of resources, supporting ongoing optimization efforts.
- **Interoperability:** Considering the heterogeneous nature of edge devices, resource profiling should address interoperability challenges. Profiling methodologies should be designed to work across diverse devices and architectures, ensuring consistency in resource characterization.

1.3.1 Task

Tasks in the context of resource management refer to the specific activities or assignments that need to be carried out in order to achieve a certain goal or objective [110]. In the case of emergency response systems, tasks can include activities such as allocating resources, coordinating rescue efforts, and managing communication channels. Mathematically, tasks can be formulated as decision variables in an optimization model [246]. For example, in the context of an emergency braking system, the objective function could be to minimize the response time for emergency vehicles to reach the accident site, while the decision variables could represent the allocation of resources to different tasks, such as dispatching ambulances, redirecting traffic, and providing medical assistance. Constraints in the model could include limitations on the availability of resources, such as the number of ambulances or the capacity of medical facilities, as well as operational constraints, such as the maximum response time allowed or the need to prioritize certain tasks over others.

A task T_v is a task generated by a specific vehicle v. This task can be described using a tuple $< \sigma_v, h_v, \tau_{max_v} >$, where, σ_v represents the size of the task data, h_v represents the amount of computation resource (measured in CPU cycles) required to execute the task, τ_{max_v} denotes the maximum delay constraint for the vehicular application that generates the task T_v.

1.3.2 Objective Function

An objective function is a mathematical formula that calculates the result used by the decision-maker to determine how well their choices work out. It is either maximized or minimized, depending on the goal of the decision-maker. For example, in a profit maximization problem, the objective is to maximize profit. The objective function is of the form

$$Z = ax + by \tag{1.1}$$

where x, y are the decision variables. The function $Z = ax + by$ is to be maximized or minimized to find the optimal solution. Here the objective function is governed by the constraints $x > 0, y > 0$. The optimization problems which need to maximize the profit, minimize the cost, or minimize the use of resources, make use of an objective function.

1.3.3 Decision Variable

A decision variable is an input value chosen by the decision maker in a problem represented by a model. It represents the values that the decision-maker can control or choose. The resources to be utilized, the time to be allocated, or the people to be deployed can all be the decision variables for an objective function. The objection function has the variables x and y as the decision variables.

- For cloud-fog offloading decisions, the binary variable represents BV_v^l (*where* to offload), i.e.,

$$BV_v^l = \begin{cases} 1, & \text{if vehicle } v \text{ is connected to location } l \\ 0, & \text{otherwise (to cloud)} \end{cases} \tag{1.2}$$

- The binary variable BV_{vf}^b represents a decision on bandwidth allocation. Generally, the SDN-controller makes an optimal decision on how much of BU b for fog node f is to be allocated to serve task $\mathcal{T}_v$. Accordingly, optimal flow-rules are placed in the SDN-enabled BSs.

$$BV_{vf}^b = \begin{cases} 1, & \text{if vehicle } v \text{ is connected to location } l \\ 0, & \text{otherwise (to cloud)} \end{cases} \tag{1.3}$$

- Binary variable BV_v^f represents the decision on vehicle-to-fog association, i.e., to indicate whether fog node f is available for $\mathcal{T}_v$. It also reflects the decision on *when* to offload task $\mathcal{T}_v$ to the other fog node f^*.

$$BV_v^f = \begin{cases} 1, & \text{vehicle } v \text{ is associated to fog node } f \\ 0, & \text{otherwise} \end{cases} \tag{1.4}$$

1.3.4 Constraints

Constraints are restrictions or requirements that decision variables must meet. They define the limitations or conditions that the decision-maker needs to consider when making decisions. Constraints can be expressed as equations, inequalities, or derivatives, depending on the problem. For example, in a linear programming problem, the constraints or the objective function are linear, while in a nonlinear programming problem, some of the constraints or the objective function can be nonlinear.

The limitations of the values that the variables x and y can take are the constraints of the objective function. The constraints of the objective function are expressed as linear inequalities in x and y. The constraints $x > 0$ and $y > 0$ are the simplest representation of non-negative restrictions. As an example, the number of boys is represented as x, and the number of girls is represented as y, and the constraint exists that the number of children who can be accommodated in the bus is less than 60. This constraint can be expressed as an inequality: $x + y < 60$.

1.3.5 Illustrative Example

In the context of resource management in vehicular fog computing, the objective function refers to the mathematical formula that calculates the result used to determine the efficiency or effectiveness of resource allocation and utilization. Decision variables determine the allocation of computing power to different fog nodes. The constraints ensure that the total energy consumption of all fog nodes does not exceed a certain threshold and that the communication bandwidth between fog nodes and vehicles meets a minimum requirement.

- **Objective function:** The objective function in resource management for vehicular fog computing can be mathematically defined as maximizing the overall resource utilization while minimizing the energy consumption. This can be represented as:

$$\text{Maximize}: \sum(\text{resource\_utilization}) - \sum(\text{energy\_consumption}) \tag{1.5}$$

- **Decision variables:** The decision variables in this context represent the allocation of resources, such as computing power, storage capacity, and communication bandwidth, to different fog nodes or vehicles. Let's consider the decision variable as X, which represents the allocation of computing power to each fog node. This can be represented as: $X = [X1, X2, X3, ..., Xn]$ Where $X1, X2, X3, ..., Xn$ represent the allocation of computing power to each fog node.

- **Constraints:** The constraints in resource management for vehicular fog computing include factors such as the availability of resources, the capacity of fog nodes or vehicles, the communication capabilities, the energy constraints, and the quality-of-service requirements. Let's consider two constraints:

- **Energy Constraint:** The total energy consumption of all fog nodes should not exceed a certain threshold. This can be represented as:

$$\sum(energy\_consumption) <= E_m ax \tag{1.6}$$

- **Bandwidth Constraint:** The communication bandwidth between fog nodes and vehicles should meet a minimum requirement. This can be represented as:

$$\sum(Communication\_bandwidth) >= B_m in \tag{1.7}$$

- **Objective function:**

$$\text{Maximize} \sum(resource\_utilization) - \sum(energy\_consumption) \tag{1.8}$$

- **Decision variables:** $X = [X1, X2, X3, ..., Xn]$ (allocation of computing power to each fog node)

  Constraints:

$$\sum(energy\_consumption) <= E_{max}(energy\_constraint) \tag{1.9}$$

$$\sum(\text{communication\_bandwidth}) >= B_{min}(\text{bandwidth\_constraint}) \tag{1.10}$$

1.4 Resource Management Sub-Problems

Resource management refers to the process of allocating and utilizing resources efficiently and effectively to achieve specific goals or objectives [153]. Efficient resource management in fog computing involves balancing these resources based on the requirements of applications, user needs, and the dynamic nature of the fog environment. This optimization contributes to improved performance, reduced

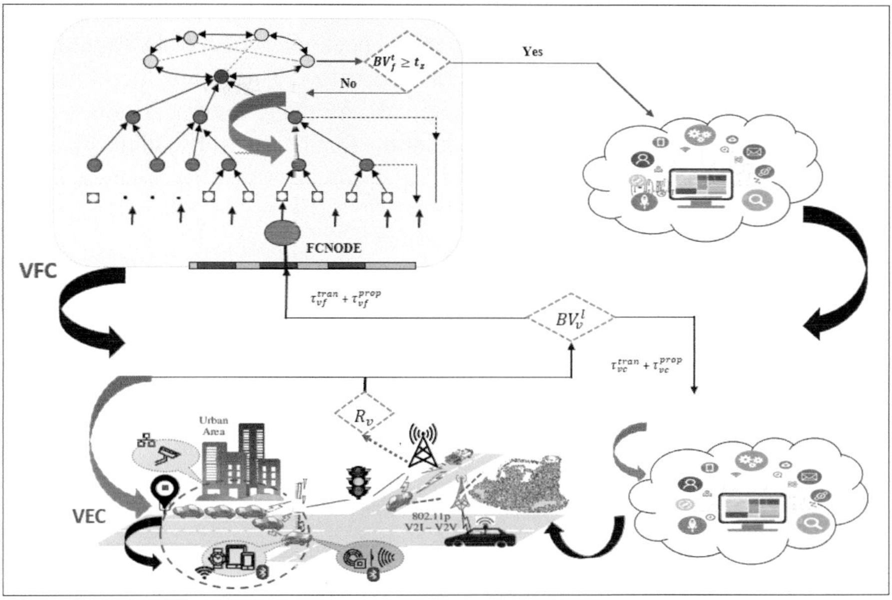

Figure 1.2: Basic topology of fog computing for vehicular network.

latency, and enhanced reliability in fog computing systems [181]. Resource management in vehicular fog computing involves balancing workloads on client devices and computation servers, placing task images on storage servers, and balancing I/O interrupt requests among storage servers [175]. It plays a crucial role in ensuring efficient and quality-of-service-aware allocation of resources. The utilization of resources in vehicular fog computing greatly impacts the overall performance and effectiveness of the system. Effective resource management schemes are essential for maximizing the utilization and minimizing the cost of cloud computing infrastructure [10]. The Figure 1.2 gives an overall view of it. These schemes aim to select vehicles with sufficient resources in an efficient and QoS-aware manner. It involves balancing workloads on client devices and computation servers, placing task images on storage servers, and balancing I/O interrupt requests among storage servers. It involves eight main fields: task offloading, application placement, resource scheduling, load balancing, resource allocation, resource provisioning, service work flow, resource orchestration, and resource pooling.

1.4.1 Task Offloading

It is a crucial aspect of resource management that involves deciding where to execute tasks in a distributed fog environment. The goal is to optimize resource utilization, reduce latency, and enhance system performance [169]. Task offloading involves dynamically distributing computing tasks from edge devices to fog nodes or the cloud based on various criteria, deciding which tasks should be processed locally and which should be offloaded to fog nodes or the cloud [259]. This problem can be formulated mathematically using optimization techniques. Here is a generic mathematical formulation for the task offloading problem in fog computing:

Binary decision variable

The task offloading problem is formulated using binary decision variables to determine the most efficient allocation of tasks to fog nodes and cloud servers. These variables, along with other parameters, are used to find the optimal solution that meets the defined objectives and constraints of the problem. In the context of task offloading, binary decision variables x_{ij} and y_{ik} can be added in the following way:

$$x_{ij} = \begin{cases} 1, & \text{if task } i \text{ is offloaded to fog node } j \\ 0, & \text{otherwise} \end{cases} \tag{1.11}$$

$$y_{ik} = \begin{cases} 1, & \text{if task } i \text{ is offloaded to cloud server } k \\ 0, & \text{otherwise} \end{cases} \tag{1.12}$$

Objective function

The optimization model aims to minimize or maximize the objective function, such as minimizing total latency and energy consumption in task offloading.

$$\begin{aligned} \text{Minimize} \quad Z = & \sum_{i\in I}\sum_{j\in J}(d_{ij}+t_{ij})\cdot x_{ij} + \sum_{i\in I}\sum_{k\in K} d_{ik}\cdot y_{ik} \\ & + \alpha \sum_{i\in I}\sum_{j\in J} E_{ij}\cdot x_{ij} \\ & + \beta \sum_{i\in I}\sum_{k\in K} E_k \cdot y_{ik}. \end{aligned} \tag{1.13}$$

Where
I = set of tasks to be offloaded, $I = 1,2,\ldots,n$.
J = set of fog nodes, $J = 1,2,\ldots,m$.
K = set of cloud servers, $K = 1,2,\ldots,p$.
d_{ij} = computation delay of task i at fog node j.
t_{ij} = transmission delay for offloading task i to fog node j.
f_j= processing capacity of fog node j. c_k = processing capacity of cloud server k.
E_{ij} = energy consumption for offloading task i to fog node j.
E_k = energy consumption for processing task i at cloud server k.

1.4.2 Resource Allocation

The process of assigning available resources to different tasks or applications in a way that maximizes overall system performance or utility is subject to constraints such as resource availability, application requirements, communication, and computational overhead [147]. This can be modeled as an optimization problem, such as linear or integer programming, where the objective is to maximize utility or performance while satisfying the constraints [175]. Mathematically, the resource allocation problem can be formulated as follows:

Decision variable

The optimization model aims to determine the optimal values of x_{ijk} to minimize the total cost while satisfying resource allocation constraints and capacity limitations.

$$x_{ijk} = \begin{cases} 1, & \text{if resource } k \text{ is allocated to application } i \text{ at fog node } j \\ 0, & \text{otherwise} \end{cases} \tag{1.14}$$

Objective Function

In this case, we aim to minimize the total cost associated with allocating resources to applications at fog nodes.

$$\text{Minimize } Z = \sum_{i \in I} \sum_{j \in J} \sum_{k \in K} c_{ijk} \cdot x_{ijk} \tag{1.15}$$

Each component can be explained as follows:
Z = the total cost, which we want to minimize.
I = set of tasks to be offloaded; $I = 1, 2, \ldots, n$.
J = set of fog nodes, $J = 1, 2, \ldots, m$.
K = Set of cloud servers, $K = 1, 2, \ldots, p$.
d_{ij} = computation delay of task i at fog node j.
c_{ijk} = the cost associated with allocating resource k to application i at fog node j.
x_{ijk} = binary decision variable.

1.4.3 Resource Scheduling

The process of determining the order and timing of resource allocation for different tasks or applications, taking into account various factors such as deadline requirements, communication and computational overhead, and energy consumption [177]. This can be modeled as a scheduling problem, such as job-shop scheduling or real-time scheduling, where the objective is to minimize the completion time or maximize the number of completed tasks while satisfying the constraints. A generic mathematical formulation for the resource scheduling problem in fog computing is given below:

Decision variable

$$r_{ij} = \begin{cases} 1, & \text{if task } i \text{ is assigned to fog node } j \\ 0, & \text{otherwise} \end{cases} \tag{1.16}$$

Here, rij is a binary decision variable that indicates whether task i is scheduled to fog node j.

Objective function

The objective function is to minimize the total completion time or energy consumption.

$$\text{Minimize } Z = \sum_{i \in T} \sum_{j \in J} t_i \cdot r_{ij} \tag{1.17}$$

Here Z is the total completion time, and the goal is to minimize it by making appropriate decisions regarding task assignments to fog nodes. T = set of tasks to be offloaded; T = 1, 2,. . . ,n.
J = set of fog nodes; J =1, 2,. . . ,m.
r_{ij} = binary decision variable.
t_i = execution time of task i.
C_i = total computation capacity of fog node j.

1.4.4 Resource Provisioning

The process of dynamically adjusting resource allocation and scheduling based on changing application demands, resource availability, and network conditions, in order to optimize system performance and resource utilization. This can be modeled as a feedback control problem, where the system adjusts the resource allocation and scheduling based on feedback from the environment.

1.4.5 Resource Pooling

It involves aggregating and managing computing, storage, and networking resources to create a shared pool of resources within the fog computing infrastructure [240]. The goal is to optimize resource utilization, enhance flexibility, and provide a scalable infrastructure by allowing resources to be dynamically allocated and de-allocated based on demand.

1.4.6 Resource Orchestration

It is the intelligent and automated coordination of resources within a fog computing environment, aiming to enhance efficiency by allocating resources based on demand, priorities, and predefined policies, while adapting to changing conditions and workload variations.

1.4.7 Service Workflow Management

The process of designing, orchestrating, and optimizing the sequence of tasks or services in an application workflow within a fog computing environment. Its purpose is to ensure efficient execution, minimize latency, improve responsiveness, and optimize resource utilization. Fog nodes coordinate to execute different components of an application workflow, ensuring task distribution based on factors like data location, processing requirements, and service dependencies.

1.4.8 Load Balancing

It is the process that distributes computing tasks across multiple fog nodes to ensure efficient resource utilization and minimize overload risks. Its primary goal is to balance computational and communication loads, prevent bottlenecks, and optimize system performance. Load balancing algorithms dynamically distribute tasks based on node loads, processing capabilities, and network conditions. Below is a generic mathematical formulation for the load balancing problem in fog computing. The annotations are listed below.
T = set of tasks to be offloaded, $T = 1, 2, \ldots, n$.
J = set of fog nodes, $J = 1, 2, \ldots, m$.
L_j = load on fog node j.
w_i = workload or computational requirement of task i.
x_{ij} = binary decision variable.

Decision Variable

In the context of the load balancing problem in fog computing, decision variables can be represented as x_{ij}.

$$l_{ij} = \begin{cases} 1, & \text{if task } i \text{ is assigned to fog node } j \\ 0, & \text{otherwise} \end{cases} \tag{1.18}$$

Objective function

The objective function represents the goal of minimizing the maximum load on any fog node.

$$\text{Minimize } Z = \max_{j \in J} L_j \tag{1.19}$$

1.4.9 Application Placement

It involves determining the optimal fog nodes on which to deploy and execute specific components or services of an application [216]. The purpose is to strategically place application components to minimize latency, meet Quality of Service (QoS) requirements, and optimize resource utilization. It can be expressed mathematically as follows:

Decision Variable

Let a_{ij} be the binary decision variable indicating whether application i is placed on fog node j. In standard form, we represent it as:

$$a_{ij} = \begin{cases} 1, & \text{if application } i \text{ is placed on fog node } j \\ 0, & \text{otherwise} \end{cases} \tag{1.20}$$

Objective Function

It can be defined as minimizing the overall latency or maximizing the overall resource utilization:

$$\text{Minimize } Z = \sum_{i \in A} \sum_{j \in F} L_{ij} \cdot a_{ij}. \tag{1.21}$$

or

$$\text{Maximize } Z = \sum_{i \in A} \sum_{j \in F} R_{ij} \cdot a_{ij}. \tag{1.22}$$

1.5 Types of Optimization Problems

An optimization problem refers to the task of finding the best solution from a set of feasible solutions, with the objective of maximizing or minimizing a certain criterion [42]. The goal is to find the optimal values of decision variables that satisfy the constraints and optimize the objective function.The different types of optimization problems are represented in the Table 1.2 and in Figure 1.3.

Different types of optimization problems include:

- Linear programming problem: In a linear programming problem, the objective function and constraints are linear. The goal is to find the optimal values of decision variables that satisfy the linear constraints and optimize the objective function.
- Nonlinear programming problem: In a nonlinear programming problem, either the objective function or the constraints (or both) are nonlinear. These problems involve finding the optimal values of decision variables that satisfy the nonlinear constraints and optimize the objective function.

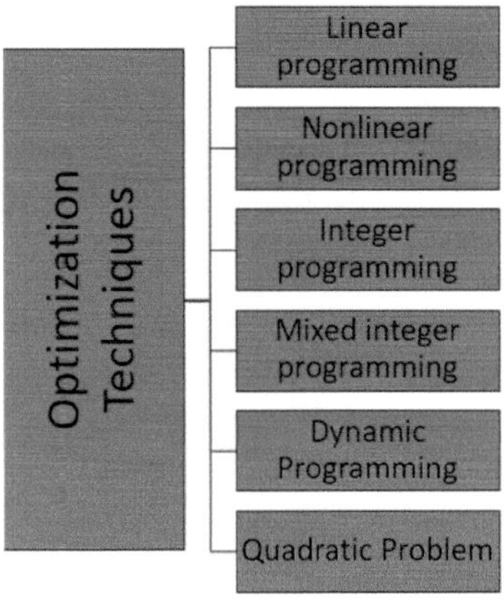

Figure 1.3: Types of optimization problem.

Table 1.2: Types of optimization Problems, LO-Linear optimization, NLP-Non-linear optimization Problems, ILP-Integer Linear optimization, MIP-Mixed integer Programming, QP-Quadratic optimization.

Properties	LO	NLP	ILP	MIP	QP	Dynamic
Objective Function	Linear	Nonlinear	Linear or Nonlinear	Linear or Nonlinear	Quadratic	Dynamic (Can Change Over Time)
Decision Variables	Continuous	Continuous	Discrete	Mix of continuous and discrete	Continuous	Continuous Over time
Constraints	Linear	Linear or Nonlinear	Linear	Linear or Nonlinear	Quadratic	Can be time-dependent
Optimality Criteria	Global minimum	Local or global Minimum	Global minimum	Global minimum	Global minimum	Over time trajectory optimal
Complexity	Generally simple	Can be complex	Generally simple	Can be complex	Can be complex	Can be complex

- Quadratic problem: A quadratic problem involves an objective function that is quadratic in nature. The decision variables may have quadratic terms, and the constraints may also involve quadratic expressions. The goal is to find the optimal values of decision variables that satisfy the quadratic constraints and optimize the quadratic objective function.
- Integer programming: Integer programming is a type of optimization problem where the decision variables are restricted to being integers. This adds an additional level of complexity compared to linear programming, as the feasible solutions must satisfy both the linear constraints and the requirement of integer values for the decision variables.

- Mixed-integer programming: Mixed-integer programming is a variation of integer programming where some decision variables are allowed to take on integer values while others can be continuous. This allows for a more flexible modeling approach, as certain variables may be better represented as continuous while others require integer values.
- Dynamic programming: Dynamic programming is a method used to solve optimization problems that can be divided into smaller subproblems. It involves breaking down the problem into smaller overlapping subproblems and solving them in a systematic manner. Dynamic programming is particularly useful for problems with overlapping substructures and optimal substructure properties.

1.6 Solving Resource Management Problems using Reduction to Standard Optimization Problems

The common mathematical formulation for resource management is often represented as an optimization problem, where decision variables are determined to optimize an objective function subject to various constraints [35]. While the specific formulation varies based on the nature of the resource management problem, a generic form can be expressed as follows:

Generic Resource Management Optimization Problem:

Decision Variables:

Let x_i represent the decision variable associated with the allocation or utilization of resource i. The nature of x_i depends on the specifics of the resource management problem, such as binary (allocated or not), continuous (quantity allocated), or integer (number of units allocated).

Objective Function:

Maximize or minimize an objective function $f(x_1, x_2, ..., x_n)$ that quantifies the overall goal of resource management. The objective function could involve maximizing efficiency, minimizing costs, or achieving a balance between conflicting objectives.

Maximize or Minimize $f(x_1, x_2, \ldots, x_n)$

Constraints: Subject to a set of constraints that reflect the limitations, requirements, and relationships within the resource management problem. Constraints are denoted by:

$g_j(x_1, x_2, \ldots, x_n)$, where j ranges from 1 to m.
$g_1(x_1, x_2, \ldots, x_n) \leq b_1$.

$g_2(x_1,x_2,\ldots,x_n) \geq b_2.$

.

.

$g_m(x_1,x_2,\ldots,x_n) = b_m.$

Here, $b_1,b_2,\ldots,b_m$ are constants, and the inequalities can be of the form $\leq,\geq$, or $=$, depending on whether the constraint is a less than or equal to, greater than or equal to, or equality constraint.

This generic formulation can be tailored to a variety of benchmark optimization problems by modifying the decision variables, objective function, and constraints to match the characteristics of each specific problem. The formulation can be customized to address specific benchmark optimization problems, offering a systematic and mathematical approach to resource management across various contexts.

In this context, we will explore how resource management problems can be addressed by mapping them to five key benchmark problems: the Knapsack Problem, Traveling Salesman Problem (TSP), Transportation Problem, Vehicular Routing Problem (VRP), and N-Queens Problem.

1.6.1 Formulation of a Resource Management Problem into a Knapsack Problem

A knapsack is a type of container or bag. In the knapsack issue, we provided certain things with their respective values and weights. We must place the goods in the bag that provide the greatest value while remaining within the bag's total stated weight. It is illustrated in Figure 1.4. There are two kinds of knapsack problems: the 0/1 knapsack problem and the fractional knapsack problem. The 0/1 knapsack problem indicates whether or not the item will be totally added to the bag. It denotes that the objects are not divisible. Because of the fractional knapsack problem, we can divide the item to compute the value/weight ratio and then place it in the bag. The Knapsack problem is a classic optimization problem in which the goal is to select a subset of items with given weights and values to maximize the total value while keeping the total weight constraint in mind. Formulating resource management as a Knapsack problem entails mapping the resource management problem's features to the Knapsack problem's components.

Mapping Resource Management to Knapsack:

- Resources as Items: Each resource corresponds to an item in the Knapsack problem.
- Resource Value: The value associated with each resource represents its importance, utility, or contribution to the overall objective.

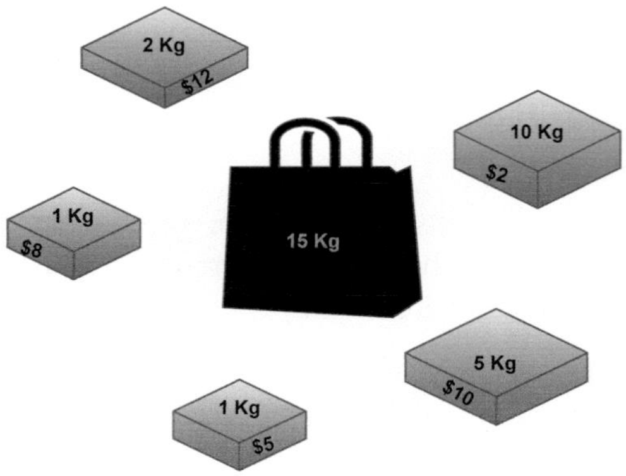

Figure 1.4: Knapsack problem.

- Resource Weight: The weight of each resource corresponds to a measure of its consumption or requirement (e.g., processing power, bandwidth, energy).
- Knapsack Capacity: The total capacity of the Knapsack represents the available resources or budget that cannot be exceeded.

Here's a generic formulation:

Decision Variables:

Let xi be a binary decision variable representing whether resource i is selected (1 if selected, 0 otherwise).

$$x_i = \begin{cases} 1, & \text{if resource } i \text{ is selected} \\ 0, & \text{otherwise} \end{cases} \tag{1.23}$$

Objective function

Maximize the total value of selected resources.

$$\text{Maximize} \sum_{i=1}^{n} (x_i \cdot \text{Value}_i) \tag{1.24}$$

Constraints

The total weight of selected resources does not exceed the capacity constraint.

$$\sum_{i=1}^{n} (x_i \cdot \text{Weight}_i) \leq \text{Capacity} \tag{1.25}$$

Where:
n is the total number of resources.
$Value_i$ is the value of resource i.
$Weight_i$ is the weight of resource i.
Capacity is the maximum allowable total weight.

1.6.2 Formulation of Resource Management Problem into a Traveling Salesman Problem

The Traveling Salesman Problem (TSP) is a well-known optimization problem in which the goal is to identify the most efficient path that visits a set of cities exactly once and returns to the starting location while minimizing total distance or tour cost. Figure 1.5 provides a representation of it.

We have a salesman who wants to visit n cities once and then return to the starting city with the least amount of trip expenditure. The trip cost c_{ij} from city i to city j is known ahead of time. A "tour" is a city-by-city itinerary that begins and ends in the same place and visits each city exactly once. The goal is to find the cheapest tour possible.

The process of incorporating a resource management problem into the Traveling Salesman Problem entails mapping the components of the resource management scenario to the elements of the TSP.

Mapping Resource Management to Traveling Salesman Problem:

- Resources as Cities: Each resource corresponds to a city in the TSP.
- Distance or Cost: The distance or cost between cities represents a measure associated with the resources, such as communication latency, cost, or any other relevant metric.

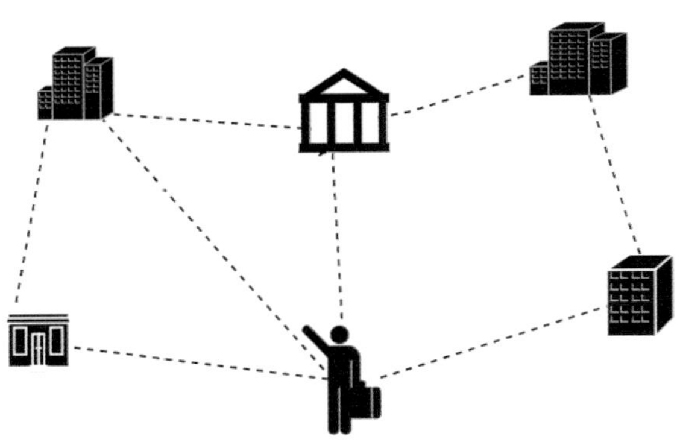

Figure 1.5: Travelling salesman problem.

- Objective: The objective is to minimize the overall "tour cost," which could represent the overall cost or efficiency of managing the resources in a specific order.
- Visit Each City Exactly Once: Ensuring that each city is visited exactly once corresponds to efficiently managing each resource in the resource management scenario.

Decision Variables:

$$x_{ij} = \begin{cases} 1, & \text{if city } j \text{ is visited immediately after city } i \\ 0, & \text{otherwise} \end{cases} \tag{1.26}$$

Objective function:

$$Min \sum_{i=1}^{n} \sum_{j=1}^{n} c_i j \cdot x_{ij} \tag{1.27}$$

Constraints:

$$x_{ij} \in [0,1] \quad \forall_{i,j} \in [1,n] \tag{1.28}$$

$$\sum_{j=1}^{n} x_{i,j} = n \quad \forall i \in [n] \tag{1.29}$$

$$\sum_{j=1}^{n} x_{i,j} = n \quad \forall j \in [n] \tag{1.30}$$

1.6.3 Formulation of Resource Management Problem into Transportation Problem

Formulating a resource management problem into a transportation problem involves mapping the components of the resource management scenario to the elements of the problem. The Figure 1.6 demonstrates the transportation problem.

Mapping Resource Management to Transportation Problem:

- Suppliers as Resource Sources: Each supplier corresponds to a source of resources, such as fog nodes or data centers.
- Consumers as Resource Destinations: Each consumer corresponds to a destination for the resources, such as end-users or applications.
- Supply and Demand: The supply at each supplier represents the available resources at that source. The demand at each consumer represents the required amount of resources at that destination.

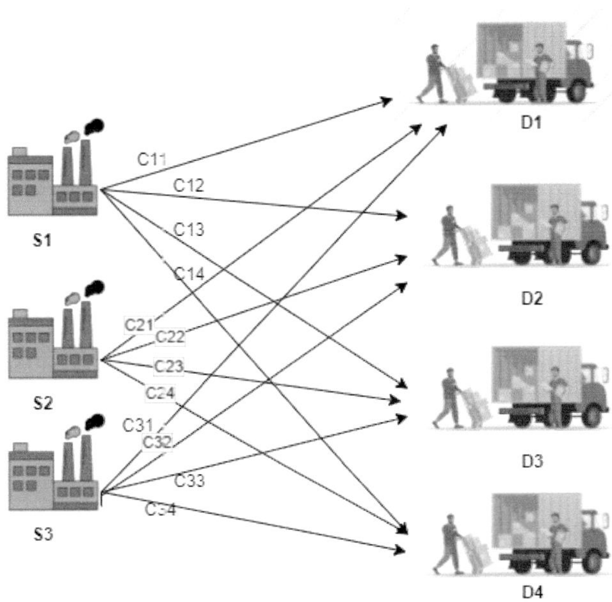

Figure 1.6: Transportation problem.

- Cost of Transportation: The cost of transporting resources from a supplier to a consumer represents the efficiency or cost associated with using that resource for a particular task.

Problem formulation

Let s_i represent a supplier i of resources and c_j represent a consumer (destination) requiring resources. We can define:
s_i as the supply at supplier i.
d_j as the demand at consumer j.
c_{ij} as the cost of transporting resources from supplier i to consumer j.

1.6.4 Formulation of Resource Management Problem into Vehicular Routing Problem

Decision Variables:

Let x_{ij} represent the amount of resources transported from supplier i to consumer j. Compared to previous problems, here the decision variables are real numbers (Not binary). Such optimization problems are called real-valued optimization Problem.

Objective Function:

The objective is to minimize the overall cost of utilizing resources to meet the demands at various destinations.

$$\text{Minimize} \sum_{k}\sum_{i}\sum_{j} c_{ij} \cdot x_{ij} \tag{1.31}$$

Constraints:

Supply Constraints:

Ensure that the total amount of resources supplied from each supplier does not exceed its capacity.

$$\sum_{j} x_{ij} \leq s_i \quad \forall i \tag{1.32}$$

Demand Constraints:

Ensure that the total number of resources demanded by each consumer is met.

$$\sum_{i} x_{ij} \geq d_j \quad \forall j \tag{1.33}$$

Vehicle routing is a challenge that requires determining the most efficient way to carry out transportation operations and deliver items using a fleet of vehicles over multiple routes. The Figure 1.7 gives a visual instance of it. The purpose of this problem is to discover the best possible set of routes for the cars to take in order to reach each stop while minimizing overall distance travelled, vehicle count, delivery time, and fuel expenses. Other aspects will be considered, such as the capacity of each vehicle, the time frame, and the service time necessary at every stop.

Mapping the aspects of the resource management scenario to the VRP's components is the process of converting a resource management problem into a vehicular routing problem. Here's a generalised version.

Mapping Resource Management to Vehicular Routing Problem:

- Vehicles as Resources: Each vehicle represents a resource capable of delivering services or managing resources.
- Locations as Service Points: Each location represents a point where resource management or service is needed.

Problem Formulation

Let v_k represent a vehicle k in the fleet and l_i represent a location (node) i that needs resource management or service.

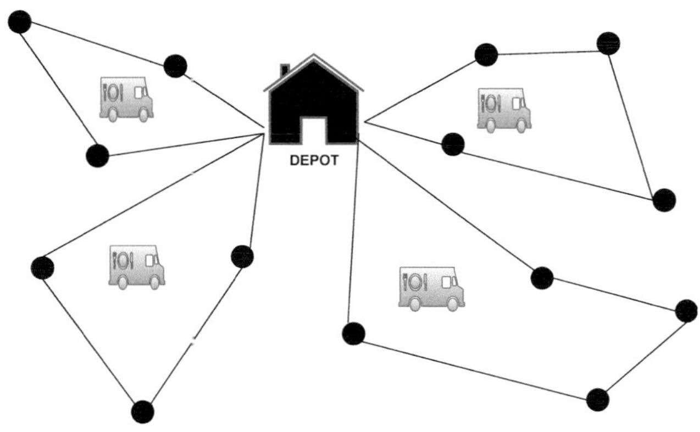

Figure 1.7: Vehicular routing problem.

Decision Variables:

The binary decision variables x_{ijk} determine whether a vehicle travels directly from one location to another.

$$x_{ijk} = \begin{cases} 1, & \text{if vehicle } k \text{ traverses directly from location } i \text{ to } j \\ 0, & \text{otherwise} \end{cases} \tag{1.34}$$

Objective Function:

Minimize the total cost associated with vehicle travel, service, or resource management.

$$\text{Minimize} \sum_k \sum_i \sum_j c_{ij} \cdot x_{ijk} \tag{1.35}$$

where
c_{ij} is the cost (distance, time, etc.) associated with traveling from location i to location j.

Constraints:

Vehicle capacity constraints:

Ensure that the total demand (resource requirement) served by a vehicle does not exceed its capacity.

$$\sum_i \sum_j d_i \cdot x_{ijk} \leq Q_k \quad \forall k \tag{1.36}$$

where
d_i is the demand (resource requirement) at location i and Q_k is the capacity of vehicle k.

Flow Conservation Constraints:

Ensure that each location is either the origin or destination for exactly one vehicle.

$$\begin{cases} 1, & \text{if } i \text{ is the origin of a vehicle} \\ -1, & \text{if } i \text{ is the destination of a vehicle} \\ 0, & \text{otherwise} \end{cases} \tag{1.37}$$

1.6.5 Formulation of Resource Management Problem into N-Queens Problem

The N-Queens problem is a classic combinatorial problem in which the goal is to arrange N chess queens on an N-Queens chessboard so that no two queens threaten each other, and it is represented in Figure 1.8. In order to convert a resource management problem into an N-Queens problem, the components of the resource management scenario must be mapped to the elements of the N-Queens problem. The solution to the puzzle is a configuration of the chessboard where:
We have exactly N queens positioned on the board.
There is exactly one queen in every row.
There is exactly one queen in every column.
Any diagonal we could draw over the board should hit, at most, one queen.

Mapping Resource Management to the N-Queens Problem:

- Queens as Resources: Each queen represents a resource or entity that needs to be placed in a specific location.
- Board as Resource Space: The chessboard represents the space or environment where resources (queens) need to be placed.

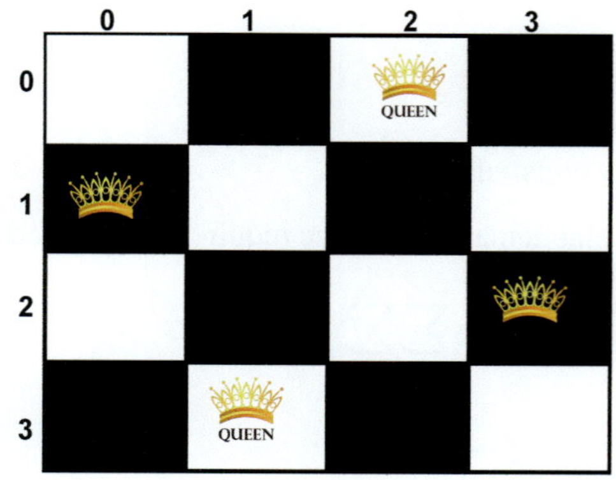

Figure 1.8: N Queens problem.

Problem Formulation

Let Q_i represent a queen i. Let B represent the N×N chessboard.

Decision Variables:

Let x_{ij} be a binary decision variable representing whether a queen is placed in cell (i, j) of the chessboard (1 if placed, 0 otherwise).

$$x_{ij} = \begin{cases} 1, & \text{if queen is placed in cell } (i, j) \text{ of the chessboard} \\ 0, & \text{otherwise} \end{cases} \tag{1.38}$$

Objective Function:

We want to maximize the number of queens placed on the board. So since our x_{ij} variables will be 1 when a queen exists in the given square and 0 otherwise, we can keep track of the number of queens on the board by just summing up all of the i, j:

$$\max f = \sum_{i=1}^{N} \sum_{j=1}^{N} x_{ij} \tag{1.39}$$

Constraints:

Queen Placement Constraints:

Ensure that each row and each column have at most one queen.

$$\sum_{i} x_{ij} \leq 1, \quad \forall j \sum_{j} x_{ij} \leq 1, \quad \forall i \tag{1.40}$$

Diagonal Constraints:

Ensure that no two queens threaten each other diagonally.

$$\sum_{k=0}^{N-1} x_{i+k,j+k} \leq 1, \quad \forall i, j \text{ such that } i+k < N \text{ and } j+k < N \tag{1.41}$$

$$\sum_{k=0}^{N-1} x_{i-k,j+k} \leq 1, \quad \forall i, j \text{ such that } i+k < N \text{ and } j-k < N \tag{1.42}$$

1.7 Leveraging Vehicular Fog Computing for Object Detection in Automatic Vehicles: A Case Study

This study delves into the application of vehicular fog computing for real-time object detection in automatic vehicles (AVs) focusing on a critical scenario involving the detection of a toddler about to cross the road. The analysis focuses on task propagation and offloading, as well as the communication flow between the AV, RSU, Access Point, and cloud server. Additionally, the study formulates computation offloading problems with the objective of optimizing processing tasks between the Av's onboard computer and the cloud server, emphasizing factors such as latency, computational resources, and network bandwidth [108].

Consider a scenario where an AV equipped with sensors detects a toddler about to cross the road in front of it. The AV's onboard computer processes the sensor data to assess the situation and recognize the need for emergency braking to prevent a potential collision. The Figure 1.9 gives a visual representation of this scenario. In response, the AV communicates with nearby roadside units (RSUs) to relay information about the detected emergency situation. The RSUs forward this data to an access point, which then transmits it over the wide area network (WAN) to a cloud server for further analysis. The cloud server, leveraging its high computational capabilities, performs advanced processing to assess the situation and determine whether emergency braking is necessary. If deemed necessary, the cloud server sends instructions back to the access point, which communicates with the AV to initiate emergency braking, thus preventing a potential collision with the toddler.

Task propagation and offloading become critical components in ensuring the timely detection of objects, such as in the scenario with a toddler crossing the

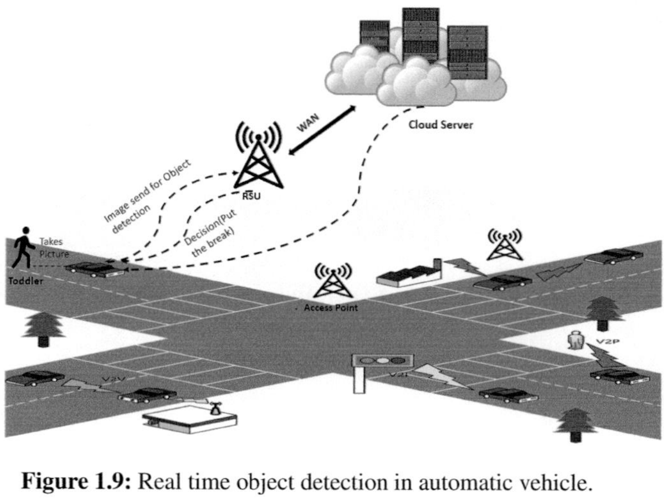

Figure 1.9: Real time object detection in automatic vehicle.

road. The computational offloading problem can be mathematically formulated as an optimization task, aiming to minimize latency and enhance computational efficiency [109]. Let T_{local} represent the local processing time by the AV's onboard computer, T_{cloud} denote the cloud processing time, C_{local} represents the computational capacity of the AV's onboard computer, and $C_c loud$ denote the computational capacity of the cloud server. The optimization objective could be expressed as:

$$\text{Minimize } \alpha \cdot T_{\text{local}} + \beta \cdot T_{\text{cloud}} \tag{1.43}$$

Subject to constraints:

$$T_{local} <= C_{local}, T_{cloud} <= C_{cloud} \tag{1.44}$$

This formulation represents a trade-off between local and cloud processing times, where α and β are weights reflecting the importance of minimizing each processing time. Additionally, the integration of machine learning algorithms for predictive resource management and containerization technologies like Docker and Kubernetes for efficient resource allocation further enriches the VFC paradigm. These advancements contribute to proactive decision-making, ensuring optimal use of computational resources and facilitating seamless communication between the AV, RSU, access point, and cloud server.

1.8 Conclusion

Resource management is a critical aspect of fog computing that requires careful consideration due to the heterogeneity of devices, dynamic workloads, and distributed architecture. Effective resource management strategies must be adaptive, considering the diverse and dynamic nature of edge devices, and they must balance QoS requirements such as latency, reliability, and bandwidth to meet the expectations of various applications. Furthermore, the dynamic nature of workloads in fog computing introduces another layer of complexity. Workloads can be highly dynamic due to factors such as varying user demands, the mobility of devices, and changing network conditions. Effective resource management must employ adaptive strategies such as load balancing, task offloading, and resource provisioning to allocate resources dynamically, ensuring optimal utilization without causing bottlenecks or under utilization. This chapter provides a comprehensive overview of how resource management can be formulated as optimization problems, including the Knapsack Problem, Traveling Salesman Problem, Transportation Problem, Vehicular Routing Problem, and N-Queens Problem. Overall, the chapter provides valuable insights into the complexities of managing resources in fog computing and offers a systematic and mathematical approach to resource management across various contexts.

Chapter 2

Swarm Intelligence in IoT and Edge Computing

Reddy Priya Madupuri, Sobin, C.C., Murali Krishna Enduri* and *Satish Anamalamudi*

Swarm intelligence plays a crucial role in enhancing the performance of IoT and edge computing. Swarm intelligence, a natural systems-inspired collective decision-making paradigm, has helped to solve most of the existing issues like channel selection, routing table optimization, and scheduling operations in IoT networks. This study discusses how swarm intelligence might improve anomaly detection, energy-efficient routing, and scalable, decentralized algorithms. IoT, edge computing, and swarm intelligence enable efficient data processing, network performance, and novel solutions to complicated issues. Swarm intelligence enhances IoT and edge computing systems, bringing new ideas and solutions for the growing environment of interconnected devices.

2.1 Introduction

Technique for metaheuristic optimization inspired by biology Swarm Intelligence (SI) imitates the behavior of other biological systems, including social insects. Strong, distributed, and easily comprehensible algorithms characterize SIs [237]. The term Internet of Things, or simply IoT, refers to the network of

Department of Computer Science and Engineering, SRM University, AP, India.

* Corresponding author: sobin.c@srmap.edu.in

physical items connected to one another by sensors, software, and other internet data-sharing technologies. Distributed computing called "edge computing" moves data processing and storage closer to data sources like Internet of Things devices. The Internet of Things runs faster and with less latency with this strategy. Internet of Things issues can be solved by swarm intelligence and edge computing which is shown in Figure 2.1. SI algorithms aid decision-making, data analysis, and IoT resource allocation. Distributed computing can use scalable and efficient SI approaches with edge computing.

Swarm Intelligence Benefits Edge Computing and the Internet of Things: SI approaches let Internet of Things devices and edge servers to share processing burden, improving efficiency and scalability. SI algorithms boost data processing and decision-making while lowering latency in Internet of Things systems, improving performance and latency. Decentralized SI algorithms are robust to interruptions, making them suited for Internet of Things systems with many devices and complex operating conditions. Swarm intelligence has been used successfully with edge computing and the Internet of Things in several examples [127]:

1. SI-based bandwidth, energy, and storage optimization in IoT systems.
2. SI methods for big IoT data processing.
3. SI algorithms provide intelligent decision-making in IoT applications including traffic control, industrial automation, and smart grids.

Swarm intelligence and Internet of Things technologies present challenges, but edge computing's distributed computing capabilities and SI algorithms' scalability, efficiency, resilience, and robustness enable practical IoT applications [65].

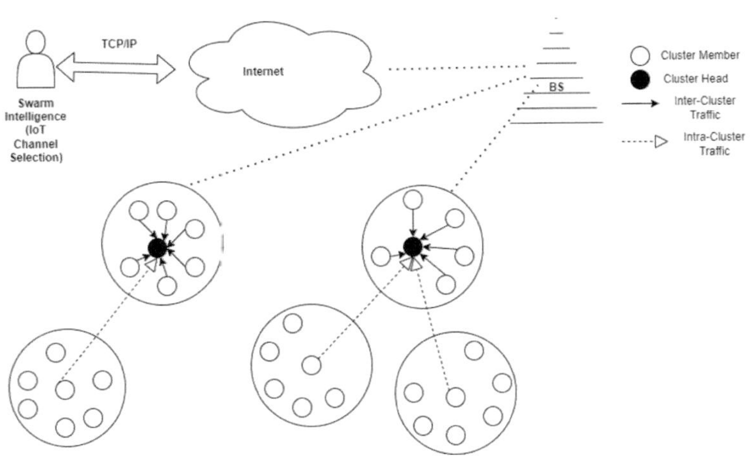

Figure 2.1: Overview of Swarm Intelligence with IoT networks.

Swarm Intelligence Applications in Edge Computing and IoT include: Smart cities use SI algorithms to improve traffic flow, energy efficiency, and public safety. SI algorithms provide intelligent systems that improve efficiency, quality, and safety in industrial automation. Smart grids use SI algorithms to optimize energy generation and distribution. SI algorithms develop intelligent healthcare systems that improve patient care and lower costs. In the future, swarm intelligence will enable more inventive and impactful IoT and edge computing applications [125].

2.2 Fundamentals of Swarm Intelligence

Swarm intelligence (SI) computationally investigates decentralized, self-organized systems. Social insects, ant colonies, bird flocks, and fish schools inspire SI algorithms. These are SI algorithm principles:

Decentralization: SI algorithms are controller less. Instead, they use individual agent interactions to attain global goals. SI algorithms self-organize to adapt to environmental changes without external intervention. SI algorithms feature emergent behavior, which is behavior that occurs from agent interactions and is not present in any particular agent [188]. Popular SI algorithms include:

Ant colony optimization (ACO): Ants' foraging behavior influenced ACO. ACO algorithms drive ants to food via pheromones. Ants that locate food strengthen the pheromone trail, which fades. ACO algorithms can discover better food sources over time.

Particle swarm optimization (PSO): Bird flocking influenced PSO. PSO methods illustrate issue solutions using particle locations and velocities. Particles exchange information in the search space. This lets PSO algorithms locate problem global optimums.

Bee colony optimization (BCO): Bees' foraging behavior influenced BCO. BCO algorithms search for food with employed, scout, and onlooker bees. Employee bees exploit food sources they have found, scout bees explore new food sources, and observer bees choose depending on employed bees' information. BCO algorithms can discover better food sources over time [34].

2.2.1 "Swarm Intelligence in IoT and Edge Computing" Principles

Swarm Intelligence is ideal for IoT and edge computing problems. SI algorithms are appropriate for distributed IoT systems with no central controller since they are decentralized. IoT systems may also adapt to environmental changes and coordinate without external intervention thanks to SI algorithms' self-organizing and emergent character.

2.2.2 Algorithms

Particle Swarm Optimization (PSO) simulates bird swarming using population-based optimization. PSO algorithms portray potential solutions with a swarm of particles. Every particle has a speed and position. Particles exchange information in the search space. This lets PSO algorithms locate problem global optimums. PSO algorithms are decentralized, self-organizing, and scalable, making them ideal for IoT and edge computing. In IoT and edge computing, PSO algorithms can optimize resource allocation, data processing, and decision-making. Particle Swarm Optimization (PSO) is based on the social behavior of bird flocks and fish schools'. PSO's main goal is to simulate particle movement around the solution space to improve a set of probable solutions. The Particle Swarm Optimization algorithm is explained in Algorithm 1.

Algorithm 1 Particle Swarm Optimization (PSO)

Require: Solution: an array representing the placement of modules on devices

Initialize particle positions and velocities
Set maximum iterations T
Set inertia weight w, cognitive weight c_1, and social weight c_2

Ensure:

Best solution found: $\mathbf{p}_{\text{best}}$
for $t = 1$ to T **do**
    for each particle i **do**
        Update velocity: $\mathbf{v}_i \leftarrow w\mathbf{v}_i + c_1 r_1(\mathbf{p}_{\text{best}} - \mathbf{x}_i) + c_2 r_2(\mathbf{g}_{\text{best}} - \mathbf{x}_i)$
        Update position: $\mathbf{x}_i \leftarrow \mathbf{x}_i + \mathbf{v}_i$
        Evaluate fitness: $f_i \leftarrow \text{fitness}(\mathbf{x}_i)$
        if $f_i < f_{\text{best}}$ **then**
            Update $\mathbf{p}_{\text{best}}$: $\mathbf{p}_{\text{best}} \leftarrow \mathbf{x}_i$
        end if
    end for
    Update $\mathbf{g}_{\text{best}}$: $\mathbf{g}_{\text{best}} \leftarrow \arg\min_i f_i$
end for

Engineering design, neural network training, and parameter tuning use PSO for complicated, high-dimensional solution spaces. The algorithm's simplicity and ability to escape local optima make it suitable for optimization problems in various domains.

Ant Colony Optimization (ACO) simulates ant foraging behavior using a population-based optimization algorithm. ACO algorithms drive ants to food via pheromones. Ants that locate food strengthen the pheromone trail, which fades. ACO algorithms can discover better food sources over time [186]. Natural optimization algorithm Ant Colony Optimization (ACO) is based on ant foraging. It excels at combinatorial optimization. The detailed pseudo code of Ant Colony Optimization is explained in Algorithm 2. ACO excels in discrete and non-linear problems. Traveling salesman, job scheduling, and network routing optimization problems have been solved using it. ACO is a reliable optimization method for complex combinatorial problems because it mimics ant foraging behavior and uses pheromones [55]. Decentralised, self-organizing, and environment-resistant ACO algorithms are ideal for IoT and edge computing systems. In IoT and edge computing, ACO algorithms can optimize routing, scheduling, and data clustering.

Algorithm 2 Ant Colony Optimization (ACO)

Require:
Initialize pheromone levels τ_{ij} for all edges (i, j)
Set maximum iterations T
Set number of ants m
Set parameters: α (pheromone influence), β (heuristic influence), ρ (evaporation rate)
Ensure:
Best tour found: $\mathbf{T}_{\text{best}}$
for $t = 1$ to T **do**
    for each ant k **do**
        Initialize empty tour $\mathbf{T}_k$
        Choose starting city randomly
        while not all cities visited **do**
            Calculate transition probabilities for unvisited cities
            Select next city based on probabilities
            Update tour and pheromone levels
        end while
        if tour length$(\mathbf{T}_k) <$ tour length$(\mathbf{T}_{\text{best}})$ **then**
            Update $\mathbf{T}_{\text{best}}$: $\mathbf{T}_{\text{best}} \leftarrow \mathbf{T}_k$
        end if
    end for
    Evaporate pheromone levels on all edges
    Deposit pheromone on edges of $\mathbf{T}_{\text{best}}$
end for

The Bee Colony Optimization (BCO) algorithm simulates bee foraging behavior using population-based optimization. BCO algorithms search for food with employed, scout, and onlooker bees. Employee bees exploit food sources they have found, scout bees explore new food sources, and observer bees choose depending on employed bees' information. BCO algorithms can discover better food sources over time. Honeybee foraging habits inspired Bee Colony Optimization (BCO), a nature-inspired optimization system. It solves optimization problems well, especially those with continuous solution spaces. See how Bee Colony Optimization works step-by-step in Algorithm 3. BCO excels in solving optimization problems with continuous and non-linear solution spaces. BCO balances local search and global exploration by emulating honeybees' cooperative foraging behavior, making it a versatile optimization strategy for complicated problems [162]. BCO algorithms are decentralized, self-organizing, and scalable, making them ideal for IoT and edge computing. In IoT and edge computing, BCO algorithms can optimize task scheduling, resource allocation, and load balancing. These are some basic IoT and edge computing swarm intelligence algorithms. Many SI methods can solve numerous problems in these sectors.

Examples of how the fundamental SI algorithms above can be applied in IoT and edge computing systems:

1. PSO: Optimizes bandwidth allocation for IoT devices in smart cities.
2. ACO: ACO optimizes mobile robot navigation in warehouses.
3. BCO may schedule tasks on a cluster of edge servers.

There are many other ways SI techniques can increase IoT and edge computing performance and efficiency. The following are some ways that swarm intelligence can be used with edge computing and IoT systems:

1. IoT resource distribution: Storage, energy, and bandwidth allocation can all be optimized with SI algorithms. SI algorithms can divide up data storage among edge servers or prioritize bandwidth for Internet of Things devices [128].
2. Large amounts of data are produced by IoT devices, which SI algorithms can distribute and process effectively. SI algorithms have the ability to move data processing to cloud or edge servers.
3. Making decisions: SI algorithms can be used by intelligent decision-making systems for Internet of Things applications such as smart grids, traffic management, and industrial automation. In a smart grid system, SI algorithms can optimize energy generation and distribution, as well as real-time traffic flow in a traffic management system.

Algorithm 3 Bee Colony Optimisation (BCO)

Require:
Initialize food source positions $\mathbf{x}_i$ for $i = 1, 2, \ldots,$ colony size
Set maximum iterations T
Set parameters: n (number of employed bees), m (number of onlooker bees), e (number of elite bees)
Ensure:
Best solution found: $\mathbf{x}_{\text{best}}$
for $t = 1$ to T **do**
    for each employed bee i **do**
        Generate a new solution $\mathbf{x}_i$ by local search around $\mathbf{x}_i$
        Evaluate fitness: $f_i \leftarrow \text{fitness}(\mathbf{x}_i)$
        if $f_i < f_{\text{best}}$ **then**
            Update $\mathbf{x}_{\text{best}}$: $\mathbf{x}_{\text{best}} \leftarrow \mathbf{x}_i$
        end if
    end for
    Select onlooker bees based on fitness
    for each onlooker bee j **do**
        Choose a random employed bee i
        Generate a new solution $\mathbf{x}_j$ by local search around $\mathbf{x}_i$
        Evaluate fitness: $f_j \leftarrow \text{fitness}(\mathbf{x}_j)$
        if $f_j < f_{\text{best}}$ **then**
            Update $\mathbf{x}_{\text{best}}$: $\mathbf{x}_{\text{best}} \leftarrow \mathbf{x}_j$
        end if
    end for
    Select elite bees based on fitness
    for each elite bee k **do**
        Generate a new solution $\mathbf{x}_k$ by global search
        Evaluate fitness: $f_k \leftarrow \text{fitness}(\mathbf{x}_k)$
        if $f_k < f_{\text{best}}$ **then**
            Update $\mathbf{x}_{\text{best}}$: $\mathbf{x}_{\text{best}} \leftarrow \mathbf{x}_k$
        end if
    end for
end for

There is hope for IoT and edge computing issues with swarm intelligence. By combining edge computing's distributed computing capabilities with swarm intelligence, we can create novel and practical Internet of Things applications.

2.3 Challenges and Opportunities in IoT and Edge Computing

2.3.1 *IoT and its Data Challenges for Swarm Intelligence in IoT and Edge Computing*

The Internet of Things, or IoT, is a network of physical objects linked to one another by software, sensors, and other technologies that allow them to exchange data online. The increasing number of IoT devices are producing massive amounts of data.

This highlights the many difficulties in swarm intelligence in IoT and edge computing, including:

1. Data Volume: IoT devices generate a lot of data, which may delay processing, analysis, and storage.
2. Internet of Things data formats have different quality and data architecture needs. Swarm intelligence systems that can handle as much data are difficult to create.
3. Real-time processing IoT data is abundant, thus swarm intelligence systems must process and analyze it quickly for decision-making [12, 164].
4. Data Security Swarm intelligence algorithms that can withstand security breaches are needed because private IoT data is easy to access.

Swarm intelligence can help edge computing and IoT overcome these issues. IoT data may be managed via swarm intelligence data filtering and compression. Eliminating and compressing unnecessary data while maintaining data integrity makes IoT data easier to process, analyze, and store, potentially reducing its vol-

Table 2.1: Challenges and Opportunities of IoT with Swarm Intelligence in different Domains.

Domain	Challenges	Opportunities
Healthcare	Privacy concerns Data security Interoperability	Patient monitoring improvement Disease prediction Remote healthcare services
Smart Cities	Network congestion Scalability issues Data management	Traffic optimization Energy-efficient infrastructure Improved public services
Manufacturing	Connectivity issues Real-time data processing Quality control	Predictive maintenance Production optimization Supply chain efficiency
Agriculture	Limited network coverage Sensor deployment Data analysis	Precision farming Crop monitoring Water resource management

ume and diversity. Swarm intelligence technologies provide real-time data processing by accurately and quickly analyzing a variety of noisy IoT data. This enables quick and decisive IoT reactions [194, 168]. For IoT data security and risk mitigation, robust swarm intelligence systems are essential. These solutions can detect and eliminate IoT data risks. Edge computing and IoT data problems can be solved via swarm intelligence. Edge computing's distributed computing capabilities and swarm intelligence algorithms' scalability, robustness, and efficiency enable new and profitable IoT applications. Edge computing and IoT swarm intelligence applications solve data problems:

1. Smart Cities: Swarm intelligence can compress and filter massive amounts of IoT data in smart cities, decreasing the cost of gathering and analyzing it and improving smart city applications.
2. Industrial Automation: Swarm intelligence analyzes IoT data in real time to identify dangers and equipment concerns for quick decision-making.
3. Healthcare: Swarm intelligence can secure highly sensitive healthcare data from IoT devices, improving data security. Swarm intelligence handles edge computing and IoT data issues well.

2.3.2 Edge Computing Constraints

The "edge computing" distributed computing paradigm is made possible by data sources, such as Internet of Things devices, being close to data processing and storage locations. Edge computing improves speed and lowers IoT latency. There are several disadvantages to edge computing, including: On edge devices, storage, CPU, and memory are usually limited. It can be difficult to execute complex swarm intelligence algorithms on edge devices. Many edge devices have limited network capacity, making data interchange with cloud servers or other edge devices difficult. Edge devices have limited batteries and resources, therefore swarm intelligence systems must save power. Swarm intelligence in edge computing is limited by resource constraints. Power and resource constraints make edge device swarm intelligence algorithm design difficult, which could affect real-time swarm intelligence in the environment of delayed data transfer.

Despite these obstacles, swarm intelligence can use edge computing. Parallel and distributed processing help swarm intelligence applications overcome power and resource constraints. With compression and aggregation, swarm intelligence algorithms can be optimized for power efficiency and edge device data flow. Edge computing restrictions can be overcome in several ways. These restrictions can be overcome to create dependable, scalable, and efficient swarm intelligence algorithms. Swarm intelligence systems on edge devices with limited resources can multitask thanks to parallel computing and distributed processing. To maximize edge device data flow and internet throughput, aggregation and compression are essential. Data compression before transmission saves network space and enhances data transfer. Swarm intelligence algorithm optimization greatly affects electricity savings. Allowing algorithms to enter low-power mode during inactivity saves electricity. Swarm intelligence techniques effectively integrate IoT and edge computing ecosystems by proactively overcoming edge computing restrictions.

2.4 Applications of Swarm Intelligence in IoT and Edge Computing

2.4.1 Energy-Efficient Routing

One of the main problems with edge computing and the Internet of Things is routing. IoT devices that run on batteries have constrained communication bandwidth. Edge computing systems are characterized by their distribution and dynamic topologies. The creation of scalable and energy-efficient routing algorithms is hampered by these considerations (refer to Figure 2.2). Energy-efficient IoT and edge computing routing methods can be developed using swarm intelligence techniques. Distributed and dynamic systems are ideal for decentralized, self-organizing swarm intelligence algorithms. Energy-saving modifications can also be made to swarm intelligence algorithms.

By imitating network nodes with a swarm of agents, swarm intelligence helps create energy-efficient routing solutions. These agents use swarm intelligence to find energy-efficient data transmission channels to improve network longevity. Swarm intelligence allocates routing responsibilities to different agent groups, resulting in energy-efficient routing algorithms for secure and efficient data transmission. Particle Swarm Optimization (PSO) routing saves energy in IoT and swarm intelligence-based edge computing wireless sensor networks. Additionally, Ant Colony Optimization (ACO) routing algorithms extend network lifespans. The "Bee Colony Optimization" (BCO) routing approach balances network latency and energy usage, making it appropriate for edge computing devices.

The combination of IoT, swarm intelligence, and edge computing allows energy-efficient routing. Swarm intelligence routing algorithms are self-organizing, scalable, decentralized, and energy-efficient, improving network efficiency.

Swarm intelligence-derived energy-efficient routing methods benefit edge computing and the Internet of Things. Swarm intelligence research uncovers new energy-saving routing alternatives for IoT and edge computing.

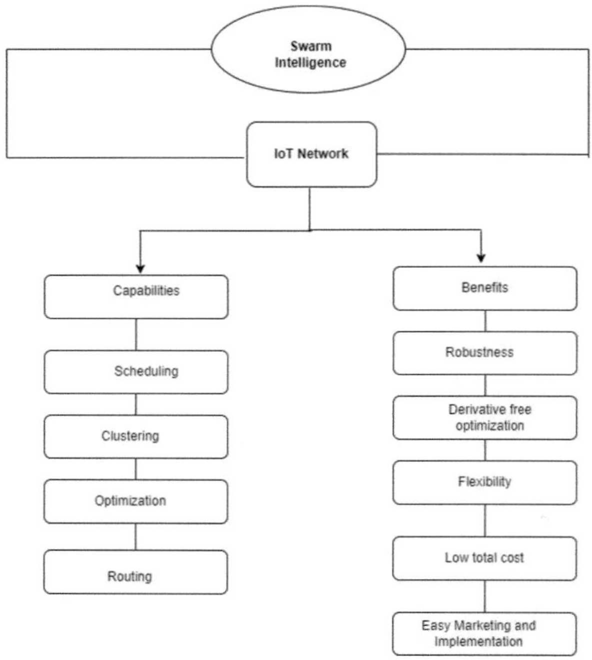

Figure 2.2: Different operations of Swarm Intelligence in IoT networks.

Natural system-inspired swarm intelligence schedules IoT operations. A complete Swarm Intelligence IoT network scheduling method includes:

1. Define the Scheduling Problem: Describe the scheduling issue in the IoT network, including duties such as data transfer, device activation, and resource allocation.

2. Represent potential schedules as solutions in the solution space, ensuring they fit with IoT device schedules.

3. Initialization: Create a swarm-like population of agents to collectively find optimal schedules, with each agent representing a potential solution.

4. Objective Function: Evaluate schedule quality, including energy efficiency, latency, and resource utilization to meet IoT network goals.

5. Local Search and Movement: Apply local search algorithms to alter schedules depending on local transmission times and device activation intervals for agent navigation in the solution space.

6. Communication and Global Information: Enable agent communication and schedule sharing, supporting swarm-wide solution sharing and exploration.

7. Track and update optimal schedules to lead the swarm towards likely solution spaces.
8. Iteration: Use swarm intelligence to improve schedule quality by using local and global data.
9. To achieve efficient solutions, monitor swarm convergence towards optimal or near-optimal scheduling.
10. Output: Optimize IoT network scheduling by retrieving the ideal swarm schedule upon convergence or meeting a preset termination threshold.
11. Implementation and Integration: Align optimal schedule with IoT network infrastructure constraints.
12. Evaluation and Adjustment: Review the IoT network schedule and alter the swarm intelligence algorithm depending on real-world performance.

Swarm Intelligence methods like PSO and ACO can be used for this. IoT network and scheduling problem characteristics determine implementation details. Finally, energy conservation and network lifespan extension are crucial. Swarm intelligence optimises edge computing and IoT routing workloads using energy-efficient routing algorithms. Swarm intelligence research will improve energy-saving routing algorithms for these complex systems.

2.4.2 RPL – Routing Protocol for IoT Networks

The Internet of Things (IoT) is a vast network of heterogeneous devices that sense and transmit environmental data. Each sensor uses multi-hop communication to send its perceived data to the sink node via its neighbors [237]. This technology is used in smart cities, industrial and building automation, disaster management, smart grid, and smart healthcare [127, 65, 125, 188]. Sensors find neighbors, form topologies, and route data. IoT systems need effective networking to handle heterogeneous traffic. An efficient routing method must consider device characteristics and limited resources. Additionally, IoT sensors confront harsh ambient conditions. Due to these devices' low energy, memory, and computational capacity, creating routing protocols is difficult [34]. A reasonable protocol goal for every pair of communicating nodes is to improve communication performance while minimizing energy usage. The lossy network, fluctuating traffic patterns, connection quality, routing loops, and convergence time must be considered to achieve this goal [92].

These Low-Power and Lossy Network (LLN) devices have been proposed with several routing protocols, including TinyAODV [257], hierarchical routing over 6LoWPAN (hilow) [186], LOAD [102], DYMO-low [77], CTP [55], and Hydro [187]. RPL, proposed by the International Engineering Task Force (IETF), ROLL

(Routing Over Low-Power and Lossy Networks) working group, is a promising routing protocol that offers IPv6 connectivity to LLN devices [162]. RPL is based on IPv6 low-power wireless personal area network (6LoWPAN) and connected to IP network via sink node (Figure 2.1) [128]. The RPL protocol has been proposed for home automation, building automation, industrial automation, urban routing, and the smart grid [12, 164]. Cisco commercially uses the RPL for FAN and CG-Mesh to enable AMI [194].

The distance vector proactive routing protocol RPL builds a tree-like destination oriented directed acyclic graph (DODAG) rooted at one or more root or sink nodes. A user-specified objective function creates directed acyclic graphs (DAGs). The OF determines the optimum route among sensor devices [194]. OF0 and MRHOF were standardized by the IETF ROLL working group as default routing metrics in RFC 6719 [168]. The OF0 selects the candidate parent node with the lowest rank in the routing tree to discover the shortest path to the sink node. The MRHOF discovers sensor node paths with the lowest link cost. The new routing path is chosen if its cost is less than the current path cost by a threshold. This is 'hysteresis.' MRHOF calculates connection quality using the anticipated transmission count (ETX) measure per the standard. ETX also analyzes link-layer congestion but not node congestion. In a heterogeneous traffic environment, picking a routing path based on the least hop count and link quality is inefficient load-balancing. Packet loss occurs near the sink node due to a high relay burden in dense networking. If the chosen parent node in RPL has numerous child nodes, the overloaded preferred parent is more likely to fail since its energy drains faster. Inefficient OFs create a routing architecture with imbalanced load and energy distribution, especially for nodes near the sink.

LLNs are used in everything from temperature monitoring to high-volume multimedia services that need fast transmission. Not using network capacity causes significant congestion and packet loss in the LLN diverse traffic environment. High relay burden, imbalanced load, and inadequate resources cause node failure. ETX-based nodes periodically broadcast probe packets to test link quality. The receiving node rebroadcasts the probing packet, increasing network congestion. If a node fails, RPL does local and global repairs. Local repair involves a child node routing packets through its sibling node or switching to the parent node. Gateways or sink nodes start global repairs. Both scenarios cause network delay and control overhead, which lowers performance. The network splits so that no path can communicate with the sink node when the number of failed nodes increases, making it non-operational [162].

Lack of regularity defines heterogeneity. Due to heterogeneous traffic load, some nodes create high traffic while others generate minimal traffic. Sensor nodes form a tree structure to send data to the root node in normal operation. Traffic is uniform and constant. This traffic pattern is predictable and load imbalance is improbable. The heterogeneous situation has a substantial load imbalance problem since each node has a different transmission interval and load.

In heterogeneous traffic load scenarios, such as Figure 2.2, where node 5 can choose node 2 or 3 as its parent node, the load imbalance may not be attributable to subtree size. Since node 3 has only one child node, node 5 may choose it as its parent. This is likely in the same traffic load condition. In this network, node 3 has a higher workload than node 2 due to node 6's high packet creation rate (60 ppm). In a heterogeneous traffic network, more routing subtree nodes do not necessarily equal more traffic flows through them. In parent nodes with big subtrees incur substantial queue losses. A parent node with a big subtree may not have high queue losses in a diverse traffic pattern. Due to unbalanced traffic generation rather than subtree size, heterogeneous traffic causes imbalanced loads and queue losses. ETX shows link-level losses, not node-level packet losses. However, hop count just shows hops, regardless of traffic load or queue losses. Also, using queue size as a decision metric does not accurately indicate load. In an uneven traffic load pattern, node burden matters. In such instances, protocol design must include latency or average end-to-end (E2E) delay and overhead as performance parameters. Therefore, a new routing metric that solves load balancing in diverse traffic load scenarios is needed.

2.4.3 Anomaly Detection

Debugging begins with identifying data points that differ from system behavior. Anomaly detection is essential for fraud detection, industrial problem diagnosis, and network security monitoring. Swarm intelligence improves anomaly detection in edge computing and the IoT. Swarm intelligence systems may quickly identify changes and infer complex conclusions from patterns. Swarm intelligence systems excel in quickly and effectively identifying anomalies in complicated datasets, improving edge computing and IoT anomaly detection processes.

Swarm intelligence lets bots represent system data points, enabling anomaly detection techniques. Swarm intelligence agents can be trained to detect data anomalies across systems. Agents should build swarm intelligence-based anomaly detection methods together. These agents use swarm intelligence to cooperate and detect abnormalities.

Several swarm intelligence methods can detect anomalies in edge computing and IoT:

1. Mutupuri et al. (2023) proposed a machine learning-based approach for malware detection in IoT networks [166], these kind of approaches can be integrated with PSO which identifies irregularities like credit card fraud.

2. Computer network attack irregularities can be detected using Ant Colony Optimization (ACO).

3. Bee Colony Optimization (BCO) finds industrial machinery malfunctions.

For edge computing and IoT anomaly detection methods, swarm intelligence can improve efficiency. Scalable, decentralized, self-organizing, and successful anomaly detection systems can be created using swarm intelligence algorithms. Additionally, swarm intelligence can be used to create anomaly detection algorithms for IoT and edge computing systems. Swarm intelligence may develop more innovative and effective anomaly detection solutions for diverse applications.

Swarm intelligence systems excel at anomaly identification for edge computing and the Internet of Things due to their robustness to aberrations and data noise. Swarm intelligence must be able to handle noisy and incomplete data from edge computing and IoT devices. Edge computing and Internet of Things systems are dynamic and unpredictable, therefore swarm intelligence algorithms must react to system changes. Due to edge computing and IoT devices' spread nature and limited processing capacity, distributed swarm intelligence algorithms must be designed. Swarm intelligence can build reliable, scalable, and adaptive anomaly detection solutions for these systems.

2.4.4 Sensor Network Management

Numerous possible uses for swarm intelligence exist to enhance sensor network management. These are a few examples:

1. Setting up the network numerous functions, including data routing, sensor placement, and parameter selection, can be automated throughout the sensor network creation process by using swarm intelligence algorithms. Thus, even when sensor network performance increases, less human intervention may be needed.

2. Gathering and filtering of data: Sensor data can be filtered by swarm intelligence systems at the network edge. This could result in less data being transferred to the cloud, saving money and enhancing network efficiency.

3. Swarm intelligence systems can discover anomalies in sensor data in real time. This facilitates the identification of the network or environmental problems—such as equipment malfunctions or environmental threats—that the sensors are keeping an eye on.

4. Fault tolerance: Fault-tolerant sensor networks can be developed by the application of swarm intelligence techniques. Swarm intelligence has the

potential to improve sensor network management in a number of ways. Here are a few illustrations:

(a) A distributed sensor network configuration protocol using particle swarm optimization (PSO).

(b) An algorithm for collecting sensor data at the network edge uses ant colony optimization (ACO).

(c) Bee colony optimization (BCO) was utilized to create a real-time sensor anomaly detection algorithm. Fault-tolerant sensor network routing uses Firefly algorithm.

(d) Swarm intelligence algorithms can create more efficient, scalable, and robust sensor networks.

Swarm intelligence can improve sensor network management in more ways besides those listed above. As swarm intelligence evolves, sensor network management solutions will become more innovative and effective.

2.5 Integration with Machine Learning and Deep Learning

2.5.1 Swarm Intelligence and Machine Learning

SI and ML are complementary technologies that can solve IoT and edge computing problems. SI algorithms optimise IoT and edge computing resource allocation, data processing, and decision-making. ML algorithms can learn complex data patterns and make predictions and judgments. In many ways, SI and ML algorithms can increase IoT and edge computing performance and efficiency. SI algorithms can distribute ML algorithm training and execution among IoT devices and edge servers. This can increase ML algorithm scalability, performance, and decision delay. Examples of SI and ML algorithms being merged to increase IoT and edge computing speed and efficiency. Federated learning using swarm intelligence algorithms: IoT devices or edge servers can train ML models using swarm intelligence algorithms. This can boost ML training privacy and scalability.

In IoT and edge computing systems, swarm intelligence algorithms may train reinforcement learning agents to complete difficult tasks. Swarm intelligence methods can train reinforcement learning agents to optimize edge computing resource distribution or operate self-driving cars. Pre-trained machine learning models can impart knowledge to new models through swarm intelligence techniques. As a result, less data and time may be required to train ML models on novel tasks. Integration of SI and ML may improve the effectiveness and performance of edge computing and IoT. Innovative and effective IoT and edge computing solutions can be achieved by combining the learning and decision-making powers of ML algorithms with the scalability, durability, and adaptability of SI algorithms.

2.5.2 Deep Learning at the Edge

Artificial neural networks are used in deep learning to understand complex data patterns. Machine translation, natural language processing, and picture classification are three areas where deep learning shines. Deep learning algorithms are computationally expensive to train and run, though. Applying deep learning techniques to IoT and edge computing systems with constrained computational resources is challenging. The solution to this issue may lie in deep learning at the edge. Edge servers and the Internet of Things host deep learning models. Performance and latency reductions are advantageous for deep learning applications. Deploying deep learning models is challenging on IoT devices and edge servers because of their limited processing capabilities. Lightweight deep learning models designed for edge servers and IoT devices can help solve these problems.

Deep learning models can be installed on edge servers and IoT devices with the aid of federated learning. Several devices can train a model using federated machine learning without sharing data. This can reduce cloud data transmission and protect data privacy. Innovative and effective edge computing and Internet of Things applications could be made possible by deep learning at the edge. We can create strong IoT and edge computing tools by integrating deep learning algorithms' learning and decision-making skills with edge computing's scalability and efficiency. Here are some edge computing and IoT instances of deep learning:

1. Smart cities: Intelligent traffic management systems and smart city object detection and tracking use deep learning.
2. Deep learning is used to construct intelligent production systems and predict machine problems in industrial automation.
3. Healthcare: Deep learning is being utilized to create sophisticated medical imaging systems to detect diseases and abnormalities.

The fast expanding field of deep learning at the edge could change how we interact with the world.

2.6 Experimental Analysis

Cooja is a network simulator which is proposed to simulate the performance of the IoT networks in a virtual environment. In general, the network is virtually created with some sample network and the performance metrics are generated through the flow of the data within the virtual network. Later on, log files will be used to test the performance of the IoT network. But, in this case, we generate the predicated performance metrics through swarm intelligence and generate the log files by using the machine and deep learning approaches. Later on, the predicated metrics are used to analyse the performance of the IoT network in terms of network throughput and end-to-end delays.

1. Goal and Importance: Simulation of Swarms The simulator simulates IoT agent swarms using Swarm Intelligence. Swarm Intelligence algorithms are tested for IoT energy efficiency, latency, and resource utilization.

2. Simulation Setting: The simulator simulates IoT devices, sensors, communication channels, and more. Realistic Example: It creates realistic IoT deployment scenarios using architecture, device density, and communication.

3. IoT Swarm Intelligence Algorithms: The simulator uses PSO, ACO, BCO. Swarm Intelligence algorithm settings can be customized to test network behavior.

4. Agent Behavior Modeling Representing Agents: Swarm agents are IoT. The simulator simulates agent decision-making using Swarm Intelligence and local and global data.

5. Swarm Intelligence's impact on IoT devices must be quantified by energy consumption and efficiency. Simulation measures data transfer delay to assess network responsiveness. Resources: Simulator tracks CPU, memory, and bandwidth.

6. Visualization, Analysis: The simulator displays agent movements, communication patterns, and algorithm convergence in IoT networks. IoT Swarm Intelligence techniques are tested via simulations.

7. Scalability Testing: The simulator evaluates Swarm Intelligence algorithms as IoT grows.

8. Network Protocol Integration: Simulates IoT network interactions with current or new protocols The simulator lets researchers test and develop IoT Swarm Intelligence algorithms. Compare IoT Swarm Intelligence Algorithms using research.

9. Modularity and Extensions: The simulator can be customised for study. Integrates and tests new Swarm Intelligence algorithms.

Last, the Network Simulator for Swarm Intelligence in IoT Networks examines and optimizes algorithms. Natural system-based collaborative decision-making boosts IoT network performance.

1. End-to-End Delay: Packet travel time. Real-time software and responsive communication necessitate low latency.
2. Successful/total Packets Transmitted Packet Delivery Ratio: High packet delivery rates suggest routing protocol reliability.
3. Data Transmission Rate: Network throughput. Significant: Higher throughput indicates routing protocol data-handling efficiency.
4. Energy Efficient Routing: Energy-efficient routing extends IoT battery life.
5. Scalability: Routing protocol network scalability. Scalability guarantees routing mechanism in big IoT deployments.
6. Routing Protocol Packet Delivery Reliability: Mission-critical apps and services need routing.
7. Uphead: Additional data or control packets for routing. Important: Lower overhead boosts network efficiency.

Performance Metrics for Scheduling Protocols:

1. End of Activity An Activity Takes Time: Reduce work completion time to launch IoT apps on time.
2. Resource Utilization Optimizes Memory, Compute, and Bandwidth Important: Task performance improves with resource efficiency.
3. Scheduled Job Completion Accuracy: IoT mission-critical apps need reliability.
4. Fair IoT Device Scheduling and Resource Distribution: Equity prevents device resource monopolies.
5. Adjustable Network/job and Scheduling Protocol: In dynamic IoT, adaptive scheduling stabilizes performance.
6. Scalable Scheduler: More jobs. Scalable scheduling grows IoT networks without hurting performance.
7. Task Conflict Avoidance: Without accidents, task scheduling is more dependable and efficient.

Swarm Intelligence Scheduling and Routing Analytics:

1. Swarm Intelligence's convergence time to an optimal or near-optimal solution: Faster convergence boosts protocol responsiveness.
2. Exploration/Exploitation Balance: Swarm Intelligence balances new and old solutions. A proper balance encourages solution space utilization.

3. Swarm Intelligence Protocol Flexibility: Durable protocols are needed in high-demand IoT contexts.

These criteria help researchers and practitioners evaluate Swarm Intelligence-based IoT Network routing and scheduling protocols for efficient and reliable IoT applications.

To validate the swarm intelligence in this chapter, we simulate the automated radio channel selection in 2.4 GHz ISM spectrum for IoT application data transmission. Figure 2.3 explains about the aggregate network throughput with respect to data rate, Figure 2.4 explains about the average end-to-end delay with respect to data rate and Figure 2.5 explains about the aggregate network throughput through number of radio channels. Based on the results in Figures 2.3, 2.4 and 2.5, swarm intelligence based radio channel selection enhances the achievable network throughput with reduced end-to-end delays.

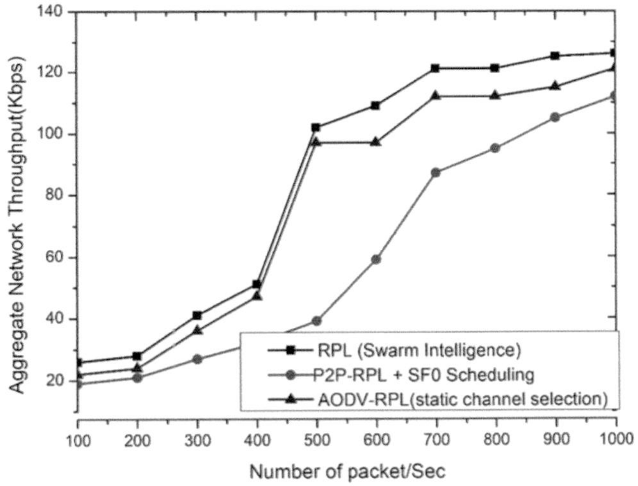

Figure 2.3: Aggregate throughput with number of data packets based on Swarm Intelligence based channel selection in IoT networks.

2.7 Future Directions

Swarm intelligence (SI) must address many issues before being deployed in practical systems. It may solve IoT and edge computing issues. SI is prone to privacy and security assaults due to its self-organizing and decentralized nature (Security and Privacy). SI algorithms handling sensitive data cause privacy issues. SI algorithms and IoT and edge computing systems employing SI must be secure and private to circumvent this. Flexible algorithms are necessary for SI to adapt to various contexts and devices. Flexible SI techniques and device and edge server

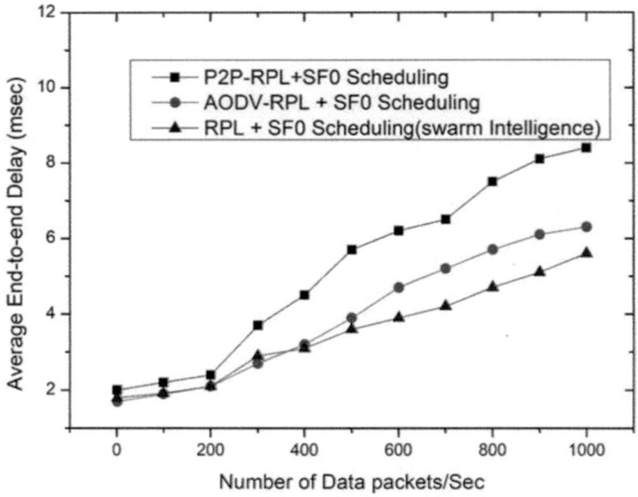

Figure 2.4: Average end-to-end delay with number of data packets based on Swarm Intelligence based channel selection in IoT networks.

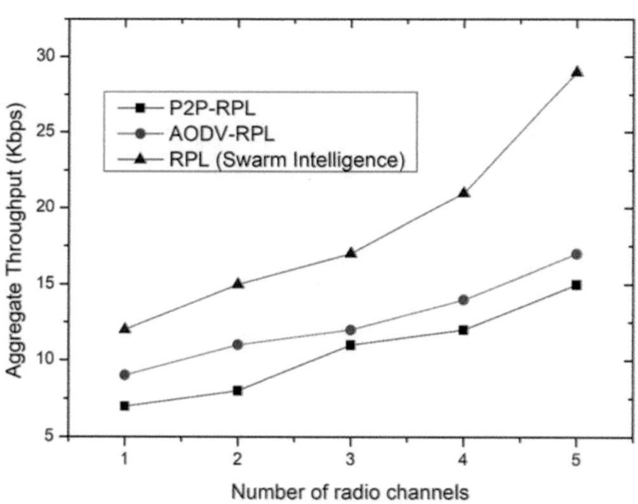

Figure 2.5: Aggregate throughput with number of radio channels through Swarm Intelligence based channel selection in IoT networks.

coordination and dissemination are needed for scalability and adaptation. SI requires flexibility to succeed in many scenarios.

Real-World Applications: SI faces deployment challenges, including functionality. SI need more research and development. Lack of industry experience limits real-world SI deployment. Overcoming deployment challenges demands reducing SI procedures for practical implementation and enhancing experts' SI understanding and application. Despite these obstacles, SI could transform edge computing and IoT. IoT and edge computing SI performance and efficiency can be improved by security and privacy, scalability and adaptability, and deployment challenges in real-world environments.

SI researchers are addressing difficulties in various areas:

1. Safe, private SI-based edge computing and IoT algorithms and methodologies.
2. Flexible and scalable SI distribution and coordination between devices and edge servers.
3. Educating industry through SI algorithm development.

SI's rapid development suggests promising adoption in edge computing and IoT systems. The Table 2.2 shows the future direction in various domains.

Table 2.2: Future directions of Swarm Intelligence in IoT and edge computing in various domains.

Domain	**Future Directions**	**Potential Applications**
Resource Management	Adaptive and context-aware allocation for dynamic workloads - Decentralized and collaborative task scheduling	Energy-efficient edge resource utilization - Optimized network bandwidth allocation
Data Analytics	Distributed in-network processing with enhanced privacy Privacy-preserving data aggregation and fusion with AI integration	Real-time anomaly detection and collaborative filtering - Enhanced recommendation systems and fraud prevention
Network Optimization	Self-organizing and self-healing networks with AI-guided swarms Collaborative spectrum sensing and allocation with optimized bandwidth usage	Dynamic routing and congestion control with improved resilience - Enhanced network scalability and adaptability
Security and Privacy	AI-assisted swarm intelligence for advanced anomaly detection Decentralized threat detection and mitigation with collaborative learning	Intrusion prevention and secure data transmission Privacy-preserving data management and access control
Smart Cities	Swarm intelligence-driven traffic management and optimization Collaborative sensor data analysis for smart waste management	Dynamic resource allocation for smart grids and utilities Optimized public transportation systems and resource distribution

2.8 Conclusion

Swarm intelligence may fix IoT and edge computing. Edge computing and IoT benefit from SI algorithms' adaptability, decentralization, and self-organization. SI algorithms optimize IoT and edge computing data processing, decision-making, and resource allocation. SI approaches allow scalable, adaptive IoT and edge computing. SI must overcome numerous challenges before being widely adopted in real-world systems. Challenges include security, privacy, scalability, adaptability, and deployments. Novel and feasible SI solutions for edge computing and IoT systems are being researched. Finally, SI may change edge computing and IoT system design and operation. Fixing SI's flaws improves edge computing and IoT systems' performance, efficiency, security, scalability, and flexibility.

Chapter 3

Optimal Deployment of Multiple IoT Applications on the Fog Computing: A Metaheuristic-Based Approach

Sai Sri Ram Kumar Macha,[1] *Pavan Kumar Chinta,*[1] *Prajwal Katakam,*[1] *Muzakkir Hussain,*[1] *Ilche Georgievski*[2] and *V. Dinesh Reddy*[1,2,*]

As more IoT devices are generating massive amounts of data today than ever before, there has been an outstanding need for solutions that can manage this data effectively. Cloud computing was a solution to this problem, but as new advancements were made in real-time data analysis and decision-making capabilities, the need for solutions with significantly reduced latency emerged. This requirement gave rise to fog computing, which introduces the challenge of placing application modules in such a way that the optimization can be maximized and latency can be further minimized to meet modern requirements. In this chapter, we propose to place these application modules utilizing the Particle Swarm Optimization

[1] SRM University-AP, Amaravati, AP, India.
[2] Department of Service Computing, IAAS, University of Stuttgart, Stuttgart, Germany.
* Corresponding author: dineshvemula@gmail.com

(PSO) algorithm. Our work also compares the results with a few other module placement algorithms. PSO algorithms explore multiple solutions and look for efficient task allocation strategies that align with the principles of social swarm behavior. Moreover, our proposed system is designed to handle fluctuating levels of resource availability present in dynamic environments, such as those offered by large fog computing infrastructures. We evaluate the performance of our system in terms of energy consumption, cost of execution on the cloud, and total network usage by making simulations in iFogSim. We observe that the application module placement using our system leads to a significant optimization of these key parameters.

3.1 Introduction

The Internet of Things (IoT) [46] has undergone significant revolutions thanks to the advancements in both hardware and software technologies. The IoT paradigm enables users to interact with the world around them in a smarter way, fostering innovation and enhancing human-machine interaction. As a result, traditional devices are being extended to a level where cities, infrastructures, health services, and more can be made smarter, effectively improving their management and quality of life. This extension has led to an exponential increase in the number of IoT devices; recent estimates suggest there are over 13 billion IoT devices globally, projected to reach around 25.4 billion by 2030.[1]

Despite these advancements, real-time analytics of IoT data continues to encounter challenges. The primary issue is the conventional practice where most of the data generated by IoT devices is transferred to and processed in the cloud. However, as the number of IoT devices proliferates, relying solely on cloud computing for data processing is not scalable and increases response times. To address this issue, fog computing has emerged as a viable approach. It acts as a bridge or layer between IoT devices and cloud services by extending cloud capabilities to the network's edge, which results in reducing latency. This approach involves the use of intermediate devices (e.g., other IoT devices, gateway devices, servers), termed fog devices or fog servers, which possess computing capabilities, such as machine learning and AI planning [85, 207]. These devices play a crucial role in determining whether to process data within the fog layer or offload it to the cloud, depending on resource availability and urgency. However, the limited resources of fog devices, such as CPU and RAM, pose a significant challenge for resource management, which is crucial for maintaining the Quality of Service (QoS) [255].

Therefore, we present a method for the module placement problem, which involves strategically assigning different application modules to various fog de-

[1] `https://www.statista.com/statistics/1183457/iot-connected-devices-worldwide/`.

vices. The assignment process considers various factors, such as power models of devices and their computational capabilities, along with constraints on module-to-device mapping. We employ the Particle Swarm Optimization (PSO) algorithm to efficiently find optimal module-to-device assignments that minimize power consumption, CPU utilization, memory usage, and total loop delay. The integration of a fog computing architecture with our PSO-based method for module placement represents a practical realization of a layered fog computing architecture, where computational tasks are effectively distributed across different layers based on devices' capabilities and the requirements of the tasks, allowing for more efficient data processing and resource utilization.

The rest of this article is organized as follows. Section 3.2 overviews related work. Section 3.3 offers relevant background. Section 3.4 introduces our PSO-based approach. Section 3.5 gives insights into the performance evaluation. Section 3.6 concludes the article.

3.2 Related Work

Optimization of energy consumption and response time has been an area where several studies and research have been conducted in recent years. Rezazadeh et al. [210] proposed a technique that leverages a simulated annealing algorithm for the placement of modules on homogeneous fog nodes, assuming a static network structure. On the other hand, Keshavarznejad et al. [131] aimed at reducing energy consumption and delays in task execution through task offloading using two meta-heuristic methods. However, their approach encounters limitations in scenarios involving a cluster of fog nodes or node failures. Furthermore, Goudarzi et al. [91] proposed an application placement technique consisting of three phases: pre-scheduling, batch application placement using a memetic algorithm, and failure recovery. This technique incorporates a weighted cost model based on execution time and energy consumption. However, this technique would require modifications to handle scenarios involving mobile IoT devices.

Mahmud et al. [151] developed a Quality of Experience (QoS)-aware application placement policy, utilizing two specific fuzzy logic models to assess and prioritize the deployment of applications based on user-defined requirements. The two modules in question are designed to handle distinct aspects of application placement within fog-enabled IoT environments. However, this dual-module approach may be considered a limitation because it restricts the evaluation to a binary framework, potentially overlooking the multifaceted nature of QoS. This limitation is particularly pertinent as it fails to incorporate a broader range of QoS attributes, such as reliability, security, and privacy, essential for a comprehensive user experience assessment in diverse IoT application scenarios. Jiang et al. [120] presented an energy-efficient offloading decision-making mechanism supported by an offloading dispatcher and a scheduler to consider response times.

Their approach is designed for environments with a set of fog devices, handling several periodic tasks, and a cloud server. However, the approach is somewhat limited as it only considers two factors of energy consumption in fog devices: computation and communication.

Jafari and Rezvani [114] proposed a joint optimization of energy consumption and time delay using a meta-heuristic algorithm. In their approach, each IoT device is connected to a fog node, which in turn is connected to a single cloud device. An observable challenge in their approach is its dependency on the fine-tuning of numerous parameters, which can be difficult to optimize. El Haber et al. [66] introduced an algorithm that focuses on meeting the latency requirements while minimizing the energy consumption of mobile devices and computational costs. This algorithm, placed on a cloud device, makes offloading decisions based on a predefined set of thresholds. This reliance on fixed thresholds may limit the adaptability of the approach to varying conditions and requirements. Apat et al. [30] introduced an approach that integrates cloud and fog computing through a middleware. However, this approach requires new strategies for mapping services to resources. A critical aspect of this study is the lack of performance analysis, which is important when adopting different techniques for service placement. The challenges associated with service placement in fog computing are multifaceted, including the difficulty of achieving optimal placement due to the dynamic nature of these environments, the limited resources available on fog nodes, security concerns, and the lack of standardization across different platforms. Guerrero et al. [93] explored the genetic optimization design of existing approaches and highlighted the need for further research to address open challenges in data management, workflow scheduling, and service placement within fog computing. Their findings indicate that there is a potential for the development of more effective designs and deployments. Particularly, there is an opportunity to work on parallel and hybrid genetic algorithms that leverage and adapt to the unique heterogeneity and distributed features of fog computing environments.

Vadde et al. [244] proposed a new approach for module placement in cloud-fog environments aimed at minimizing energy consumption. Their approach generates a random walk based on a probability distribution and facilitates a balanced compromise between the utilization of the fog landscape and the energy demands associated with running applications within that fog landscape. While their approach shows a potential for improving energy efficiency in fog computing environments, its effectiveness depends on several factors. These include the size of the fog landscape, the number of modules being placed, and the specific application being run. Natesha et al. [171] introduced a first-fit decreasing heuristic-based approach for placing IoT application modules on fog-cloud environments. While this approach presents a promising strategy, it works under a specific assumption: smaller application modules are hosted by fog nodes, whereas larger modules are forwarded to the cloud. However, this assumption may not universally apply as the optimality of module placement can vary significantly depending on the unique requirements of each specific application.

Hassan et al. [100] proposed an approach to service placement in fog-cloud environments using two algorithms. One algorithm is designed to prioritize response time, particularly for critical services, thereby ensuring efficient performance in time-sensitive applications. The other algorithm is designed to minimize energy consumption in standard services, thereby providing support for sustainability. This dual-algorithm approach offers a balanced solution that manages the trade-off between response time and energy consumption. However, this approach assumes homogeneity in the QoS requirements. Additionally, it is based on a restrictive assumption that IoT applications consist of two services, which limits the approach's applicability across more complex IoT applications. Misra and Saha [159] presented an algorithm that handles task offloading in a software-defined fog network. This network architecture conceptualized as a directed graph, consists of IoT devices, access points, and fog nodes. Given that the number of fog nodes is smaller than access points, the algorithm uses multi-hop task offloading, where each IoT device has a singular computational task at any given time. The proposed approach optimizes the paths for offloading requests, thereby ensuring efficient usage of the available fog computing resources. This approach assumes that tasks cannot be further divided, i.e., a particular task can be offloaded to only one fog node.

3.3 Background

3.3.1 The Architecture of Fog Computing

Fog computing offers a significant advancement over traditional cloud-based architectures, addressing the ever-increasing need for low-latency, high-bandwidth, and geographically distributed computing. By extending cloud computing capabilities to the network edge, fog computing enables computation, storage, and services to be distributed in a highly decentralized manner closer to the end-user devices.

The architecture of fog computing is hierarchical and is often described in these three layers:

- **Devices Layer:** This layer comprises various end-user devices, which can include a variety of IoT devices, such as sensors, actuators, and other smart devices with varying computational capabilities and power models.
- **Fog Layer:** At this layer lies a collection of cutting-edge computational tools, namely, fog nodes or edge devices that can perform complex computations at an impressive scale. These fog nodes enable local data pro-

cessing within their respective layers. As such, they comprise a range of networking tools, such as routers and switches, with advanced abilities for computation and storage. By implementing these revolutionary tools into operations, unnecessary data transmission to the cloud is reduced significantly, resulting in controlled bandwidth usage accompanied by improved response times.

- **Cloud Layer:** This layer hosts the traditional cloud infrastructure with massive computational and storage resources. Tasks that are not time-sensitive or require significant resources are performed in this layer.

The connection between these layers provides the framework for executing different tasks at the most appropriate level, balancing the need for processing speed, energy efficiency, and bandwidth usage.

3.3.2 iFogSim

Our evaluation of applications and services in a fog environment leverages the iFogSim [95] simulator. This open-source tool provides excellent capability for modeling and simulating fog computing environments. Its realistic representation of a fog computing environment makes it ideal for our research as it enables us to perform a comprehensive analysis of various design choices. The components that make up the fog environment in iFogSim include fog nodes, applications, and an interconnected topology that facilitates their interaction. We have designed two simulations: a virtual Reality (VR) game and Two Applications.

The "EEG Tractor Beam Game" is a VR game uniquely designed for brain-computer interaction, making it an ideal candidate for research in fog computing. Players, equipped with EEG headsets, participate in a mental concentration challenge where they control a game object with their brainwaves. The game operates in real-time, processing EEG signals to determine each player's concentration level. This is visualized on-screen with players surrounding a target object, competing to exert an attractive force on it through mental focus.

What makes this game particularly suitable for fog computing is its inherent need for real-time data processing near the source, i.e., the EEG headsets and smartphones, combined with a requirement for broader network interaction. Fog computing offers a solution that balances local processing demands with the expansive capabilities of cloud computing. This balance is crucial for the game's simultaneous needs of immediate EEG data processing and global multiplayer connectivity.

The game's architecture comprises three primary modules: the Client, the Concentration Calculator, and the Coordinator. The Client module manages local EEG signal processing and updates the player's game state. The Concentration

Calculator evaluates the EEG data to quantify the player's concentration level. The Coordinator functions globally, orchestrating the game among various players in different locations and continuously synchronizing each player's game state.

For modeling in iFogSim, a prominent toolkit for simulating IoT and fog computing environments, the game's application model is structured around these three modules. Each module is represented by the AppModule class in iFogSim. The data dependencies between these modules are depicted using the AppEdge class. This setup accurately reflects the real-world interactions and data flow in the game, providing a comprehensive framework for simulation. Additionally, the game involves a sensor (EEG) and an actuator (DISPLAY), further integrating it with the IoT paradigm. The control loop, central to the game's operation, is represented in iFogSim using the AppLoop class, ensuring a realistic simulation of the game's functionality and suitability for fog computing research.

In the next part of this section, we delve deeper into the specifics of the fog environment setup by discussing the environments used for simulation. These environments were made using the iFogSim Toolkit.

3.3.2.1 Virtual Reality (VR) Game

The fog computing model decentralizes the cloud by providing computing, storage, and networking services between end devices and traditional cloud servers, aiming to improve efficiency and reduce the amount of data that needs to be transported to the cloud for processing, analysis, and storage.

In the context of this VR game, fog computing becomes particularly valuable due to the high demand for real-time, low-latency processing and data analytics. Figure 3.1 presents the application model of the VR game and two applications.

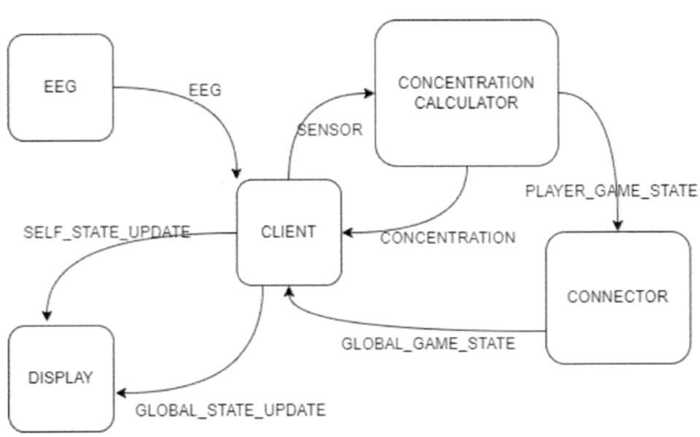

Figure 3.1: Application model of the VR Game and Two Applications.

Table 3.1: Characteristics of Fog Devices in VR Game.

Fog Device	MIPS	RAM	Uplink Bandwidth	Downlink Bandwidth	Level	Rate per Tuple	Busy Power (W)	Idle Power (W)	Uplink Latency
Cloud	44800	40000	100	10000	0	0.01	16*103	16*83.25	-
Proxy Server	2800	4000	10000	10000	1	0	107.339	83.4333	100
Gateway	2800	4000	10000	10000	2	0	107.339	83.4333	4
Mobile	1000	1000	10000	270	3	0	87.53	82.44	2

Topology of VR Game As shown in Figure 3.2, the hierarchy of the fog nodes is structured with the cloud at the apex (level 0), followed by the proxy server (level 1), department gateways (level 2), and mobile devices (level 3). This hierarchical structure helps manage resources and data processing in a distributed fashion, from edge devices (mobile devices) to the cloud.

Fog Devices In the given VR game, the fog nodes are represented by instances of the *FogDevice* class. These devices are central to the fog environment, providing the computational resources that execute various application modules. Each fog device is characterized by its processing power (MIPS), RAM, uplink and downlink bandwidths, hierarchy level in the fog, cost rate per MIPS, and power consumption characteristics (busy power, idle power). The fog devices also have specific roles, such as cloud, proxy servers, gateways (representing departments), and mobile devices (representing end-user devices in each department). Table 3.1 shows the specifications and properties of the fog nodes used in this environment.

Application The application in this environment is the VR game represented by an instance of the *Application* class. It comprises several modules, namely *client*, *concentration_calculator*, and *connector*, that communicate with each other via defined edges. The application handles tasks such as data acquisition from sensors (EEG data), calculation of player concentration levels, and communication with the global game state. The application also interacts with actuators to display game states.

AppEdges of Modules in the Application *AppEdges* represent the communication channels between the application modules, sensors, and actuators. They are characterized by their source, destination, tuple CPU length, tuple network length, tuple type, direction (UP sensor to the module, or DOWN module to module or actuator), and edge type (SENSOR, MODULE, ACTUATOR). These AppEdges define how data moves and is processed within the application.

The VR game is built as an edge application distributed across several layers of fog nodes, from mobile devices to the cloud. The application processes data from EEG sensors, calculates player concentration levels, communicates with the global game state, and interacts with display actuators. This environment allows the VR game to benefit from the reduced latency, enhanced real-time processing, and efficient resource usage provided by the fog computing model. Figure 3.2 illustrates the topology of the VR game and two applications.

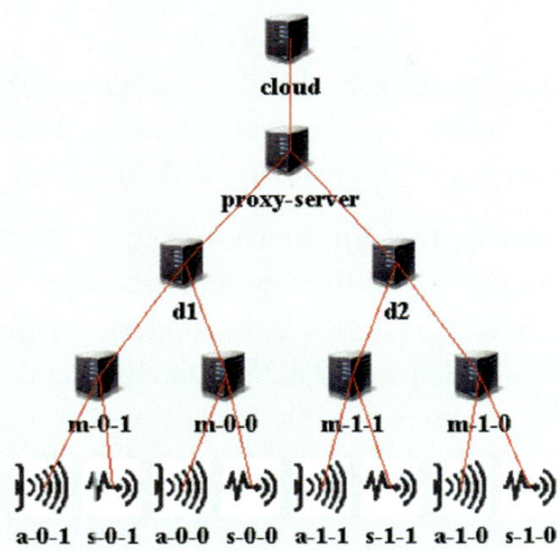

Figure 3.2: Topology of the VR Game and Two Applications.

3.3.2.2 Two Applications

The VR game influences this fog environment. The fog environment consists of multiple instances of the VR game. Each application includes several modules interconnected to form a directed acyclic graph (DAG). The modules are 'client', 'concentration_calculator', and 'connector'. Each module has its computing requirement specified as 10 MIPS. The application model also includes edges representing the flow of tuples (data) between different modules, sensors, and actuators. The properties of these edges include tuple processing time, tuple data size, and tuple type.

The topology of this fog environment is defined as the same as the topology used for VR game simulation.

Fog Devices These are the devices in the fog computing system, which can range from cloud servers to edge devices like smartphones. They are represented in the iFogsim simulator by the *FogDevice* class. Each fog device is defined by properties like hierarchy MIPS (Million Instructions Per Second), RAM, bandwidth, and latency. The fog devices in this environment include:

- *Cloud:* The top-level fog device at the hierarchy level 0. It has higher computing resources compared to other nodes.
- *Proxy Server:* This fog device is at hierarchy level 1 and serves as the parent for the gateways. The latency of connection from the Proxy Server to the Cloud is defined as 100 ms.

- *Gateways:* These devices are connected to the Proxy Server. The connection latency between gateways and the Proxy Server is defined as 4 ms.
- *Mobiles:* These are the end devices that are connected to gateways and serve as the parent for sensors and actuators. The connection latency between the smartphone and gateway is defined as 2 ms.

AppEdges of Modules in the Application The edges represent the data flow between different components of the application. Here, the term 'Tuple.UP' represents the data flow direction from sensors to modules or lower-level modules to upper-level modules, and 'Tuple.DOWN' indicates the data flow from a module to an upper-level or lower-level module to sensors. For instance, in application0, the 'EEG' sensor is connected to the 'client' module, the 'client' is connected to the 'concentration_calculator', and the 'client' is also connected to the 'DISPLAY' actuator.

Sensors and Actuators Sensors generate data (tuples), and actuators perform actions based on the data they receive. In this fog environment, two types of sensors and actuators are defined, one for each application. They are connected to mobile devices (edge devices), and their latencies are defined. The sensors follow a deterministic distribution for data transmission times.

3.4 Proposed Method

3.4.1 Particle Swarm Optimization Algorithm

Particle Swarm Optimization (PSO) is a bio-inspired optimization algorithm applied here to solve the problem of module placement in fog computing. The aim is to optimally assign application modules to fog devices in such a manner as to minimize network latency and maximize the efficiency of resource usage. The PSO algorithm utilized in this work is a swarm intelligence method that mimics the social behavior of birds flocking or fish schooling. The algorithm is initialized with a group of random solutions, known as particles, and then searches for optimal solutions by iteratively updating generations. Each particle updates its position based on its own experience and the experience of neighboring particles. This behavior helps the algorithm converge to the global best solution.

In the PSO algorithm, each particle represents a potential solution to the module placement problem, i.e., a specific assignment of an application module to fog devices. The position of a particle in the search space corresponds to the module-device assignment represented by the particle, and the fitness of a particle corresponds to the quality of its module-device assignment, as evaluated by a predefined fitness function.

3.4.2 Module Placement

The central methodology used for this project revolves around Particle Swarm Optimization (PSO), a powerful computational method that finds the optimal solution by mimicking social behaviors observed in nature. The main advantage of this algorithm is its capability to search a wide solution space quickly and efficiently, making it ideal for solving complex optimization problems like module placement in fog computing.

3.4.2.1 Fitness Function

The fitness function serves as the guiding metric that helps the PSO algorithm in its search for optimal solutions. This function evaluates the performance of a solution, with each solution represented as an integer array that maps modules to devices. The function calculates various parameters such as power consumption, CPU utilization, memory usage, and total loop delay based on the given solution. It then uses these parameters to compute the fitness value for the solution. The fitness function takes the form of the following weighted sum equation,

$$\begin{aligned} FitnessValue = W_1 * powerConsumption + & \quad (3.1) \\ W_2 * cpuUtilization + & \\ W_3 * memoryUsage + & \\ W_4 * totalLoopDelay & \quad (3.2) \end{aligned}$$

where W_1, W_2, W_3, and W_4 are weight constants that can be adjusted to prioritize different parameters according to the user's needs.

3.4.2.2 Power Consumption

In terms of power consumption, a common approach is to estimate the total energy consumed by the devices in executing the assigned modules. This can be computed as the sum of the power consumed by each device, which in turn can be calculated as the product of the device's power coefficient, CPU utilization, and memory usage. Mathematically, this can be represented as,

$$powerConsumption = \sum_{i=0}^{n} \text{powerCoefficient} * \text{cpuUtilization\_i} * \text{memoryUsage\_i} \quad (3.3)$$

for all devices i, where *powerCoefficient* is a constant representing the amount of power consumed per unit of CPU utilization and memory usage, *cpuUtilization* is the proportion of the device's CPU capacity that is used in executing the assigned modules, and *memoryUsage* is the proportion of the device's memory capacity that is used in executing the assigned modules.

Algorithm 4 Fitness Function Calculation

```
Require: solution: an array representing the placement of modules on devices
Initialize total power consumption, CPU utilization, memory usage, and loop delay to zero
Create an empty mapping of module placements
for each module-device pair in the solution do
    if the module is not allowed to be placed on the device then
        Penalize the solution with a high cost and exit
    end if
    Calculate and add the module's power consumption to the total
    Calculate and add the module's CPU utilization to the total
    Add the module's memory usage to the total
    Record the placement of the module in the mapping
end for
Calculate the total delay for all application loops based on module placements
for each application loop do
    for each module in the loop do
        if the loop's module sequence is not respected in the placements then
            Penalize the solution with a high cost and exit
        end if
    end for
end for
Calculate the overall fitness from the total power consumption, CPU utilization, memory usage, and loop delay.
return the calculated fitness
```

3.4.2.3 CPU Utilization and Memory Usage

The CPU utilization and memory usage of a device can be calculated based on the resource requirements of the modules assigned to the device. For instance, if a module requires a certain amount of CPU cycles and memory to execute, the proportion of the device's CPU and memory capacities used by the module can be calculated as,

$$cpuUtilization = \sum_{j=0}^{n} \left(\frac{cpuRequirement_j}{cpuCapacity_j} \right) \tag{3.4}$$

$$memoryUtilization = \sum_{j=0}^{n} \left(\frac{memoryRequirement_j}{memoryCapacity_j} \right) \tag{3.5}$$

for all modules *j,* assigned to the device *i,* where *cpuRequirement* and *memoryRequirement* are the amounts of CPU and memory required by a module, and *cpuCapacity* and *memoryCapacity* are the total CPU and memory capacities of the device.

3.4.2.4 Total Loop Delay

The total loop delay is a measure of the total time taken for a loop of modules to execute on the assigned devices. This can be computed as the sum of the execution times of the modules in the loop, plus the communication times between consecutive modules. Assuming that the execution time of a module depends on the device's CPU utilization and memory usage and that the communication time between two modules depends on the network latency between the devices on which they are placed, the total loop delay can be calculated as,

$$\text{totalLoopDelay} = \sum_{j=0}^{n} (\text{executionTime}_j + \text{communicationTime}_j) \tag{3.6}$$

3.5 Performance Evaluation

The primary objective of the simulation is to validate the efficacy of the newly developed ModulePlacementPSO algorithm by comparing it with two distinct module placement strategies: ModulePlacementEdgewards and ModulePlacementOnlyCloud. Each of these algorithms was subjected to a series of tests within an identical simulated fog environment using iFogSim2, ensuring a fair and controlled comparison.

ModulePlacementEdgewards represents an edge-centric placement strategy. It focuses on strategically positioning application modules across a network of fog devices. The uniqueness of this algorithm lies in its method of generating and analyzing 'leaf-to-root' paths, which are potential data routes from edge devices to the central cloud. This approach enables the algorithm to make informed decisions about module placements, ensuring that each module is hosted on a device with adequate resources and optimal network positioning, thereby maximizing overall system efficiency while respecting module inter-dependencies. ModulePlacementOnlyCloud, in contrast, adopts a cloud-centric model. It centralizes all computational tasks at a single cloud server, bypassing the intermediate fog nodes. This algorithm stands out for its simplicity and direct approach, delegating all processing to a resource-rich cloud server. It is particularly beneficial in scenarios where centralized data processing is a priority, or in cases where the deployment of additional fog nodes is impractical or not cost-effective.

To comprehensively assess the performance of these algorithms, each was tested multiple times under varying conditions. The simulations were conducted iteratively, with the number of fog nodes progressively increasing to encompass a wide range of scenarios. This approach allowed for a detailed analysis of how each algorithm performs under different network topologies and loads, highlighting the advantages and trade-offs of the ModulePlacementPSO algorithm in comparison to ModulePlacementEdgewards and ModulePlacementOnlyCloud. The results from these simulations are pivotal in demonstrating the practical applicability and benefits of the ModulePlacementPSO algorithm within the fog computing paradigm, providing insights into its potential for real-world deployments.

3.5.1 Execution Time

Execution time refers to the time required to complete the execution of the application. From the results, it's evident that the ModulePlacementPSO algorithm had an execution time of 611 seconds. In comparison, ModulePlacementEdgewards and ModulePlacementOnlyCloud took 348 and 356 seconds, respectively. Though PSO took slightly longer, it's crucial to remember that the primary aim of fog computing is to optimize resources and reduce latencies, not necessarily to reduce execution time.

3.5.2 Application Loop Delays

The loop delay measures the time a loop of modules takes to execute and transfer data back to the user. The ModulePlacementPSO algorithm has shown a loop delay of approximately 331 seconds. This is significantly more than the loop delays observed in ModulePlacementEdgewards (approx. 226 seconds) and ModulePlacementOnlyCloud (approx. 226 seconds). This suggests that there might be an area for further optimization of the PSO algorithm regarding loop delay.

3.5.3 CPU Execution Delay

The CPU execution delay refers to the time the CPU needs to execute a specific operation. Observing the CPU execution delays of particular tasks, the ModulePlacementPSO algorithm exhibits slightly higher delays than the others. While this might appear as a disadvantage, it's worth noting that the PSO algorithm aims to balance utilization, power consumption, and other factors, not merely minimize execution delay.

3.5.4 Energy Consumption

This is where the proposed ModulePlacementPSO algorithm truly shines. The total energy consumed by the fog devices in ModulePlacementPSO is noticeably less than that in the other two algorithms. ModulePlacementPSO's total energy consumption was approximately 4923517.35 units in VR Game and 4785573.73 units in Two Apps. In comparison, the total energy consumption in ModulePlace-

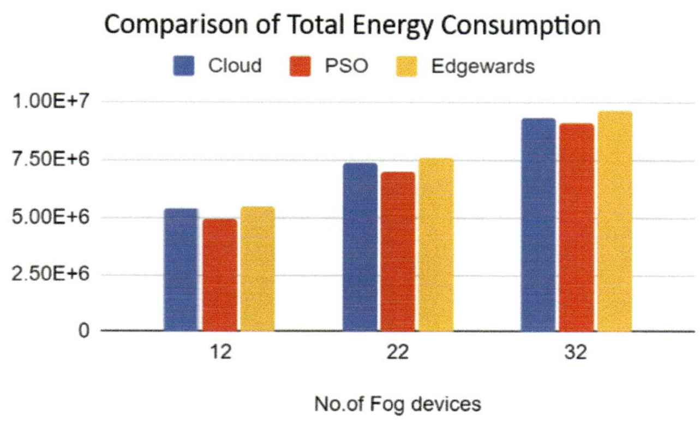

Figure 3.3: Total energy consumption in VR Game.

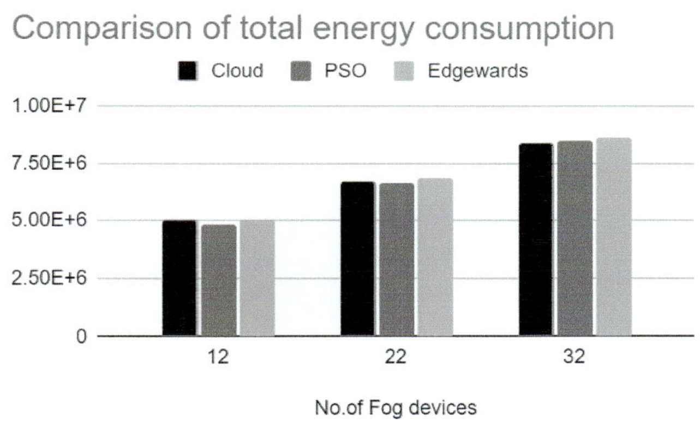

Figure 3.4: Total energy consumption in Two Apps.

mentEdgewards was about 5483899.822 units and 5039133.685 units in Two Apps, and in ModulePlacementOnlyCloud, it was around 5379382.759 units and 5025955.621 units in Two Apps. Total energy consumption in VR game and Two apps are presented in Figures 3.3 and 3.4.

Thus, the PSO algorithm outperforms the other two by reducing the total energy consumption significantly, which is an essential factor considering the need for energy-efficient solutions in the IoT environment.

3.5.5 Cost of Execution in the Cloud

The cost of execution in the cloud for ModulePlacementPSO is much less than the other two placement algorithms. For ModulePlacementPSO, the cost is approximately 11,799 units for VRGame and 317,223.99 units for Two Apps. In

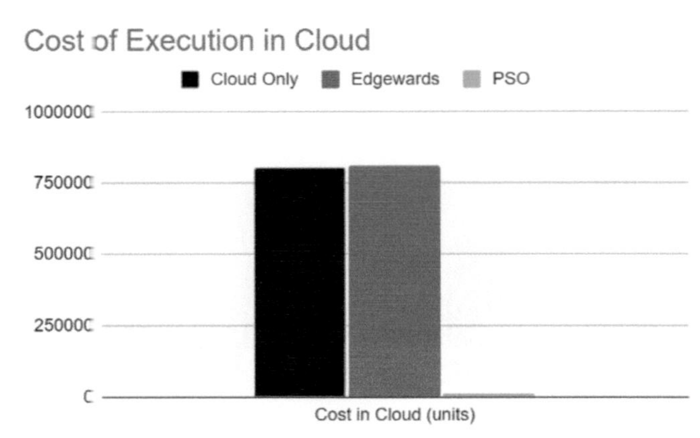

Figure 3.5: Cloud execution cost VR Game.

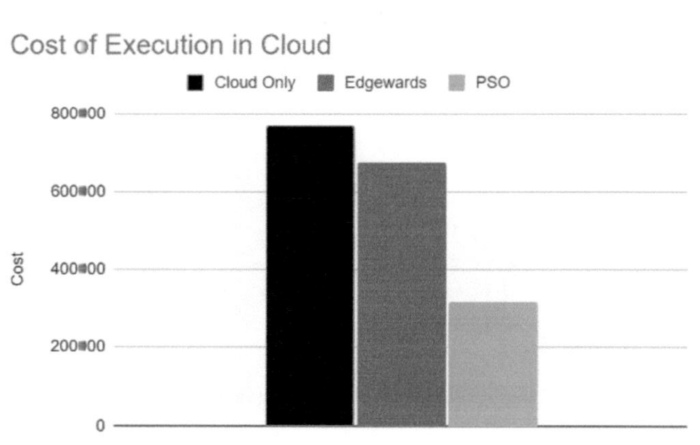

Figure 3.6: Cloud execution cost Two Apps.

contrast, ModulePlacementEdgewards has a cost of about 811,507 units in the VR Game and 675599.16 units in Two Apps, and ModulePlacementOnlyCloud has a cost of around 802,406 units in VR Game and 768852.81 units in Two Apps. Cloud execution cost of the VR game and two apps are presented in Figures 3.5 and 3.6.

These numbers suggest that the PSO algorithm has a clear advantage in minimizing cloud execution costs.

3.5.6 Network Usage

The total network usage was highest for ModulePlacementOnlyCloud approximately 208,599 units in VR Game and 284878 units in Two Apps and lowest

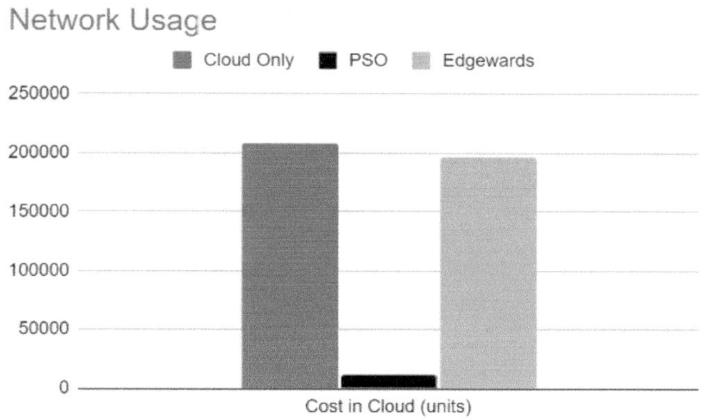

Figure 3.7: Network usage in VR Game.

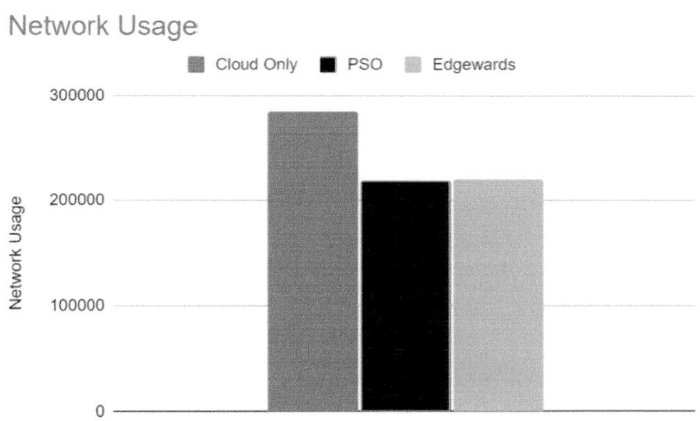

Figure 3.8: Network usage in Two Apps.

for ModulePlacementPSO approximately 11,272 units in VR Game and 219431 units in Two Apps. ModulePlacementEdgewards fell in the middle with around 196,652 units of network usage in VR Games and around 219340 in Two Apps. Network usage in VR game and Two apps are shown in Figures 3.7 and 3.8.

Therefore, the PSO algorithm seems to be a more network-efficient solution, leading to lower bandwidth consumption. Simulation Results for VR Game Fog Environment and simulation Results for Two Apps Fog Environment are presented in Table 3.2 and Table 3.3.

Table 3.2: Simulation results for VR Game Fog environment.

Metric	Cloud-Only	PSO	Edgewards
Cloud Energy	3229982.959	2672322.65	3236402.016
Proxy-Server Energy	166866.6	166866.6	166866.6
d-0 Energy	166866.6	166866.6	166866.6
m-0-0 Energy	164880	175059.491	174789.721
m-0-1 Energy	164880	175059.491	174780.113
m-0-2 Energy	164880	175059.491	174779.938
m-0-3 Energy	164880	175059.491	174593.0943
m-0-4 Energy	164880	175059.491	174647.0585
d-1 Energy	166866.6	166866.6	166866.6
m-1-0 Energy	164880	175059.491	174433.6653
m-1-1 Energy	164880	175059.491	174789.721
m-1-2 Energy	164880	175059.491	174713.1992
m-1-3 Energy	164880	175059.491	174581.7741
m-1-4 Energy	164880	175059.491	174789.721
Application Loop Delay	225.532806	330.7462827	226.4422907
Execution Time	356	611	348
Cloud Execution Cost	802406.22	11799.2	811506.656
Total Network Usage	208598.5	11272	196651.5

Table 3.3: Simulation results for Two Apps Fog environment.

Metric	Cloud Only	PSO	Edgewards
Execution Time	576	536	560
Application0 Loop Delay	224.5455085	271.7170849	273.09308
Application1 Loop Delay	227.0728193	355.1297265	354.6832769
Cloud Energy	3206315.821	2887756.21	3140538.693
Proxy-server Energy	166866.6	166866.6	166866.6
d-0 Energy	166866.6	166866.6	166866.6
m-0-0 Energy	164880	174747.7489	174746.0793
m-0-1 Energy	164880	174774.4026	174784.631
m-0-2 Energy	164880	174642.7943	174704.5793
m-0-3 Energy	164880	174639.433	174745.9368
d-1 Energy	166866.6	166866.6	166866.6
m-1-0 Energy	164880	174751.0351	174775.357
m-1-1 Energy	164880	174646.8065	174733.0693
m-1-2 Energy	164880	174480.618	174780.8848
m-1-3 Energy	164880	174534.8774	174724.6543
Cost of Execution in Cloud	768852.8091	317223.9943	675599.16
Total Network Usage	284878.5	219431	220748

3.6 Concluding Remarks

This research explores a module placement algorithm in fog computing using Particle Swarm Optimization (PSO), executed in the iFogSim environment. The findings show significant advantages, particularly in power usage, by distributing modules across fog nodes, leading to enhanced cost efficiency, reduced network congestion, and minimized energy consumption. The paper suggests investigating the PSO-based algorithm's scalability on larger fog environments and enhancing optimization with dynamic factors. Additionally, the applicability of other nature-inspired optimization algorithms and the impact of varied deployment scenarios should be explored. In summary, our research effectively demonstrates the benefits of a PSO-based module placement algorithm in fog computing, optimizing costs, network usage, and energy, thus having important implications for fog computing systems' practical implementation.

3.7 Future Work

In this research, we made use of one of the meta-heuristic algorithms, i.e., the Particle Swarm Optimization algorithm to place the application modules on fog nodes that are heterogeneous or homogeneous in such a way that. Particle Swarm Optimization algorithm to place application modules on fog nodes that are heterogeneous or homogeneous in such a way that total energy consumption, cost of execution on a cloud device, and the total network usage are optimized. In future work, the application loop delay can be further reduced and should address scalability challenges, incorporate dynamic factors, and explore alternative optimization algorithms to further enhance module placement strategies in fog computing.

Chapter 4

Machine Learning and Deep Learning with Swarm Algorithms

Mutupuri Shiva Shankar, Sruthi Sivarajan, Anil Carie, Satish Anamalamudi and *Sobin, C.C.*

The emerging disciplines of machine learning and deep learning have brought about an innovative period, restructuring domains such as image identification, natural language comprehension, and autonomous systems. However, there are still complex optimisation and decision-making difficulties that need to be addressed. Swarm algorithms, inspired by the coordinated movement of natural swarms, are emerging as innovative solutions in this field. This study investigates the productive convergence of machine learning, deep learning, and swarm algorithms, with the goal of unleashing their combined potential. We explore the core principles and mechanisms of swarm algorithms such as particle swarm optimisation, ant colony optimisation, bee-inspired algorithms, firefly algorithms, and bat algorithms. These decentralised algorithms, inspired by nature, are highly effective in addressing various optimisation goals by imitating the self-organizing behaviours observed in natural systems. Additionally, we explore the complex incorporation of swarm algorithms into the operations of machine learning. We aim to utilise swarm-based techniques to enhance the performance and generalizabil-

Department of Computer Science and Engineering, SRM University, AP, India.
* Corresponding author: satish.a@srmap.edu.in

ity of machine learning models by optimising hyperparameter tweaking, model selection, and feature engineering. This chapter examines the use of swarm intelligence in the emerging subject of neural architecture search, which is crucial for automating the complex process of designing deep neural networks. In order to strengthen our comprehension, we undertake a journey through captivating instances and tangible implementations in fields like robotics and healthcare. This emerging discipline combines the collective intelligence of groups of organisms with the data-driven capabilities of machine learning to create innovative solutions that can improve decision-making, optimise intricate systems, and drive progress in the field of artificial intelligence.

4.1 Introduction

The recent years have witnessed an unprecedented surge in Artificial Intelligence (AI) driven by the advancements in Machine Learning (ML) and Deep Learning (DL). However, amidst this data-driven revolution, a new paradigm is emerging: Swarm Intelligence. Inspired by the collective wisdom of natural swarms, swarm algorithms offer decentralized and self-organizing solutions to complex optimization and decision-making problems (see Figure 4.1). This begs the intriguing question: What happens when we fuse the data-driven capabilities of ML or DL with the self-organizing principles of swarm algorithms.

This integration marks a groundbreaking shift in perspective, where the strengths of each domain synergistically amplify the other. Imagine harnessing the data-driven insights of ML/DL to guide the exploration and exploitation of swarm algorithms, resulting in solutions that are not only intelligent but also adaptive,

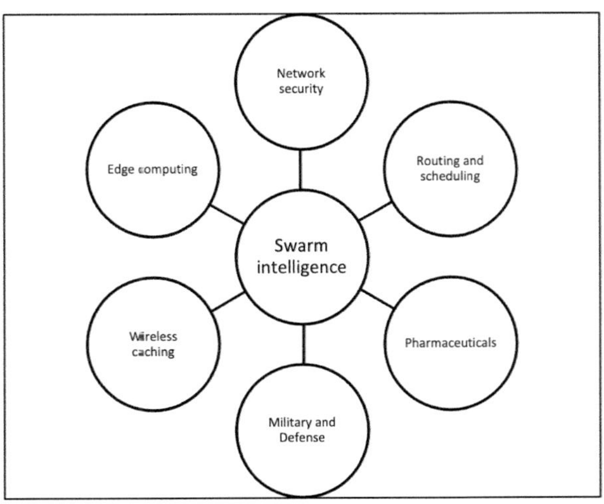

Figure 4.1: Applications of Swarm intelligence in various areas.

much like natural swarms navigating their environment. To unlock this transformative potential, we embark on a journey into the fascinating world of swarm algorithms. We begin by demystifying the core principles of traditional algorithms like Particle Swarm Optimization (PSO) and Ant Colony Optimization (ACO), before venturing into the intriguing realm of bee-inspired algorithms. This foundational understanding paves the way for seamlessly integrating these algorithms into ML/DL workflows. As we delve deeper, we witness the profound impact of swarm algorithms in optimizing ML models. They act as master sculptors, shaping and refining these models through hyperparameter tuning, model selection, and feature engineering. Furthermore, we explore the intricate dance between swarm algorithms and Deep Learning, where they automate the creation of neural architectures through Neural Architecture Search (NAS) and fine-tune the training process of deep neural networks. The transformative power of this synergy is not just theoretical. We embark on a voyage through real-world examples drawn from diverse fields like robotics, healthcare, finance, and telecommunications. Each case study unveils the remarkable adaptability of swarm algorithms in tackling practical challenges, showcasing their tangible impact on various domains.

We must acknowledge the challenges and limitations associated with this nascent field. Issues like scalability, convergence, and ethical considerations are thoroughly examined, offering a balanced perspective on the current state of affairs. Finally, we peer into the future, highlighting potential research avenues and exciting prospects in this interdisciplinary arena. Our aim is to spark the imagination of researchers, practitioners, and enthusiasts alike, inviting them to contribute to the advancement of this nascent field by exploring its untapped potential.

4.2 Swarm Algorithms Fundamentals

Imagine a colony of ants, each a tiny architect, collectively constructing towering nests. Or consider a murmuration of starlings, swirling in mesmerizing patterns across the sky. These marvels of nature reveal the power of the collective, a power harnessed by swarm algorithms. By delving into the core principles of swarm intelligence, we'll explore how these algorithms tap into the wisdom of interconnectedness, solving complex problems through a symphony of simple interactions.

4.2.1 Introduction to Swarm Intelligence

Drawing inspiration from the intricate societies of social insects, swarm intelligence mimics their collective problem-solving abilities. By empowering individual agents to interact and collaborate, this AI paradigm fosters self-organization, adaptability, and resilience, making it adept at tackling challenging optimization and decision-making tasks across diverse domains.

4.2.2 Particle Swarm Optimization

Particle Swarm Optimization is a strategy for optimization that draws inspiration from the coordinated behavior observed in bird flocks and fish schools, emulating nature's guidance. First introduced by Dr. Eberhart and Dr. Kennedy in 1995 [237], PSO has gained recognition for its straightforward yet effective approach to tackling diverse optimization challenges. In this algorithm, potential solutions are represented as particles navigating a search space. Each particle adjusts its position based on its individual experience and the collective knowledge of the entire group. Throughout the optimization process, particles persistently move toward optimal solutions, making iterative adjustments to their velocities and locations.

In each iteration, swarm particles adapt their velocities by considering their personal best positions in the past and the best position discovered by any particle in the entire swarm. The collective interaction, incorporating inertia and acceleration coefficients, propels the group toward advantageous areas within the search space. Particle Swarm Optimization (PSO) has found successful applications in various domains, including engineering design, data clustering, and neural network training. Its parallelizability and effectiveness in solving nonlinear and multi-modal optimization problems make it a versatile tool in the realm of computational intelligence. Despite its notable accomplishments, PSO encounters challenges in striking a harmonious balance between exploration and exploitation. Consequently, researchers persist in exploring diverse variations and hybrid methods to enhance its effectiveness in diverse scenarios.

4.2.3 The Ant Colony Optimization

The Ant Colony Optimization algorithm represents a bio-inspired method for optimization, drawing inspiration from the foraging behavior observed in ants. Marco Dorigo conceptualized and developed this algorithm in the late 1990s [127], showcasing its effectiveness in solving combinatorial optimization problems extensively. The fundamental concept behind ACO stems from the foraging habits of real ants, utilizing pheromone deposition on the terrain to identify the most efficient route between their nest and a food source.

In the optimization field, it is customary to depict potential solutions as paths traversing a graph. Artificial ants, as they navigate these paths, dynamically adjust pheromone levels. The probability of selecting a specific path depends on the concentration of pheromone, showing a preference for paths with higher levels. Over time, ants consistently favor shorter paths, leading to an increased concentration of pheromones along these routes. This phenomenon facilitates subsequent ants in choosing these advantageous pathways. Through iterative exploration and exploitation, the ACO algorithm gradually converges towards an optimal solution for the given problem.

ACO's adaptability spans various problem domains, making it a powerful tool for addressing combinatorial optimization challenges like the traveling salesman problem and job scheduling. Its decentralized nature and reliance on localized interactions make it highly compatible with parallel and distributed computing environments. Additionally, ACO has been extended and hybridized with other optimization techniques, showcasing its adaptability in complex problem domains. While widely recognized for its success, ACO faces challenges in scalability and convergence speed. Current research focuses on enhancing its efficacy and adaptability, particularly for broader challenges. In summary, Ant Colony Optimization stands as a testament to the effectiveness of bio-inspired algorithms, addressing complex optimization problems by incorporating principles derived from observations in the natural world.

4.2.4 Bee-Inspired Algorithms

Bee-Influenced Algorithms (BIAs) represent an engaging subset of swarm intelligence, a broader field encompassing diverse algorithms inspired by the collective behaviors observed in honeybee colonies [186]. The algorithms explored in this study aim to emulate the cooperative foraging and decision-making behaviors demonstrated by bees, proving highly valuable in addressing intricate optimization problems across diverse domains. A noteworthy instance is the Bee Algorithm, a computational approach leveraging communication strategies observed in bees during their exploration and exploitation phases of foraging activities. By facilitating information exchange among individual bees, the algorithm effectively achieves a balanced equilibrium between exploring potential solutions and exploiting promising ones. Consequently, it demonstrates an enhanced ability to discover optimal solutions across a varied range of problem spaces.

The Bee Algorithm is a renowned optimization technique involving the participation of various bee types, namely scout bees, employed bees, and onlooker bees. Each bee type plays distinct roles within the algorithm to streamline the optimization process. Scouts embark on an exploratory journey within the solution space to unveil novel possibilities, while employed bees leverage previously identified solutions. Onlooker bees, on the other hand, base their decisions on the information conveyed by employed bees regarding solution quality. The algorithm's dynamic adaptability to changing environments and problem characteristics stems from a decentralized approach, drawing inspiration from the self-organization observed in bee colonies. Bee-Influenced Algorithms have garnered significant attention due to their outstanding ability to effectively navigate intricate solution spaces. These algorithms have successfully found applications in various domains, including job scheduling, route optimization, and function optimization, showcasing their versatility and effectiveness in real-world scenarios. Table 4.1 shows the comparison of PSO, ACO and Bee-inspired algorithms.

Table 4.1: Comparison of various criteria of PSO, ACO, and Bee-inspired algorithms.

Criteria	PSO	ACO	Bee-inspired
Convergence Speed	Generally faster for continuous problems.	Slower convergence, but robust to dynamic environments.	Moderate convergence speed, often better than ACO for continuous problems.
Solution Quality	Can get stuck in local optima, potentially leading to suboptimal solutions.	Prone to finding good, but not necessarily the best, solutions.	Often finds high-quality solutions, but depends on proper parameter tuning.
Parameter Sensitivity	Requires fewer parameters, but can be sensitive to their values.	More parameters, but less sensitive to individual values.	Parameter sensitivity varies depending on the specific variant.
Computational Cost	Relatively low.	Moderate, potentially higher for complex problems.	Moderate, similar to ACO on average.

4.2.5 Other Swarm Algorithms

The scope of collective intelligence experiences a substantial expansion through the incorporation of diverse swarm algorithms that go beyond familiar paradigms like PSO and ACO. An exemplary computational technique is the Firefly Algorithm, drawing inspiration from the synchronized flashing behavior observed in fireflies. This algorithm utilizes fireflies as potential solutions for optimization problems, assessing each firefly's attractiveness based on the objective function's value. Mirroring the inherent behavior of fireflies, the algorithm prompts fireflies to gravitate towards brighter counterparts, reflecting the optimization process as solutions progressively approach heightened fitness values [65]. This algorithm's efficacy has been demonstrated across a broad spectrum of optimization tasks, spanning domains such as image processing, feature selection, and financial modeling. The Algorithm 5 shows the psuedo code for the firefly algorithm.

In the obscure realms of optimization, a distinctive algorithm arises, influenced by the murmurs of the night. The Bat Algorithm, conceived by Xin-She Yang in 2010, draws inspiration from the intricate sonar harmonies of bats as they navigate the world through echolocation [125]. Transforming bats into prospec-

Algorithm 5 Firefly Algorithm

```
Initialize a population of fireflies with random positions and light intensities
Define the objective function to be optimized
Set algorithm parameters: attractiveness, absorption coefficient, randomization parameter
while termination criterion not met do
    for each firefly i do
        for each firefly j brighter than i do
            Calculate the distance r_ij between fireflies i and j
            Calculate the attractiveness β(r_ij) using the distance and absorption coefficient
            Move firefly i towards firefly j based on attractiveness and randomization
        end for
        Evaluate the new position of firefly i using the objective function
        Update the light intensity of firefly i based on its fitness
    end for
    Rank the fireflies based on their light intensities
    Update the best solution found so far
end while
Return the best solution found
```

tive solutions, their vocalizations evolve into a rhythmic exploration, charting unknown territories within problem spaces. Similar to nimble navigators, they adjust the frequencies and volumes of their calls, creating a subtle equilibrium between convergence and diversity. This bat algorithm has demonstrated its efficacy in challenging optimization terrains, resilient against noise and dynamic obstacles. Algorithm 6 shows the psuedo code for the bat algorithm.

Understanding the soul of swarm algorithms is like mastering the whispers of a hidden language. Beneath the surface of their simple interactions lies a potent flexibility and power, ready to tackle problems that once left us stumped. But to truly appreciate this dance of collective brilliance, we must first delve into the core principles that ignite their spark. This chapter embarks on a journey of exploration, where we'll witness how these foundational ideas intertwine with the cutting-edge tapestry of machine learning and deep learning. Through this synergy, we'll witness the emergence of novel approaches that transcend conventional boundaries, pushing the frontiers of problem-solving to new heights.

Algorithm 6 Bat Algorithm

Initialize a population of bats with random positions, velocities, and frequencies
Define the objective function to be optimized
Set algorithm parameters: loudness, pulse rate, frequency minimum and maximum
while termination criterion not met **do**
  for each bat i **do**
    Generate new frequency f_i within the allowed range
    Calculate the velocity v_i using frequency and random walk
    Update the position x_i using velocity
    if random number $<$ bat i's pulse rate **then**
      Generate a local solution around x_i using a local search method
      if local solution is better than x_i **then**
        Update x_i with the local solution
      end if
    end if
    Evaluate the fitness of x_i using the objective function
    Update the loudness and pulse rate of bat i based on its fitness and iteration
  end for
  Rank the bats based on their fitness
  Update the best solution found so far
end while
Return the best solution found

4.3 Integrating Swarm Algorithms with Machine Learning

Fine-tuning the levers of machine learning models can be delicate, demanding precision and patience. These algorithms offering new optimizing parameters, seamlessly blending with existing workflows and alleviating the burden of hyperparameter tinkering [55]. This chapter delves into revealing how swarm intelligence can unlock the full potential of machine learning models, while streamlining the journey to peak performance.

4.3.1 Swarm Based Hyperparameter Tuning

Swarm strategies, exemplified by Particle Swarm Optimization and Ant Colony Optimization, present a decentralized and autonomously organizing strategy for traversing through parameter spaces. This technique proves beneficial, especially in the domain of tuning hyperparameters, aiming to pinpoint the optimal combination of model parameters. Hyperparameter tuning algorithms based on swarm principles harness the combined wisdom of a collective to adeptly traverse the intricate and expensive parameter space characterized by numerous dimensions. The result is swift convergence and enhancement in the overall model performance.

4.3.2 Model Selection and Feature Engineering with Swarm Algorithms

The tradition of model selection often involves exhaustive forays into vast landscapes of hyperparameters, which is demanding for both researcher and computer. Here, we introduce a tango partner: swarm-based optimization. Algorithms like Particle Swarm Optimization and Ant Colony Optimization bring a decentralized grace to traversing these complex settings. Table 4.2 shows the comparison of PSO and ACO on various criteria with respect to model selection and feature engineering. Instead of a doing solo, imagine a coordinated swarm, their collective intelligence guiding the search for the most harmonious model configurations [188].

In the era of machine learning, Feature Engineering plays a pivotal role. It's the art of selecting and sculpting the data's raw notes, transforming them into melodies that resonate with the model. Swarm algorithms shine here, acting as

Table 4.2: Comparison of PSO and ACO for model selection and feature engineering.

Criteria	PSO	ACO
Model Selection	Can be used for direct model selection by evaluating fitness of different configurations using the objective function.	Requires adaptation for model selection, often using information gain or similar metrics within the algorithm.
Feature Engineering	Can be used for feature selection and ranking by evaluating individual features' impact on solution quality.	Can be adapted for feature selection by analyzing pheromone trails and ant movements, but less intuitive than PSO.
Strengths	Fast convergence, especially for continuous problems. Can handle continuous and discrete features.	Robust to dynamic environments and complex feature spaces. Offers natural parallelism for distributed computing.
Weaknesses	Prone to getting stuck in local optima. Requires parameter tuning for optimal performance.	Slower convergence than PSO. Primarily suited for discrete feature spaces. Can be computationally expensive for complex problems.
Overall Suitability	Well-suited for problems with fast fitness evaluation and continuous features. May require hybrid approaches for effective feature engineering.	Effective for dynamic problems and discrete feature spaces. Requires adaptation for model selection and may be computationally expensive.

skilled conductors, identifying features that sing the loudest [34]. By exploring combinations of features with the agility of a beehive, they not only lighten the manual labor of feature selection but also ensure the model captures the data's hidden harmonies. The seamless integration of Swarm Algorithms with model selection and feature engineering creates a transformative way in the machine learning. This fusion leads to streamlined solutions around complex problems with newfound capability and adaptability. It's not just a change of tempo, it's a fundamental shift in the way we understand and utilize the power of machine learning.

4.3.3 Optimizing Training Processes with Swarm Intelligence

In the realm of model selection, conventional methods often entail exhaustive exploration of hyperparameter spaces, a task that proves both laborious and computationally demanding. This chapter introduces swarm-based optimization approaches, specifically Particle Swarm Optimization and Ant Colony Optimization. These innovative techniques provide a decentralized and efficient means of navigating intricate hyperparameter settings. Leveraging the collective intelligence of swarms, these algorithms streamline the search process, facilitating the identification of optimal model configurations.

The significance of Feature Engineering in machine learning cannot be overstated; it involves the meticulous selection and transformation of relevant features to enhance model accuracy. Swarm algorithms play a pivotal role in automating this process by discerning crucial traits within a dataset. Employing a swarm-based methodology allows for the exploration of feature subsets, guiding the algorithm towards meaningful combinations. This not only reduces the manual effort required for feature selection but also enhances the model's capability to capture significant patterns inherent in the data. The amalgamation of Swarm Algorithms with model selection and feature engineering underscores their potential to streamline and enhance the machine learning pipeline. This integration paves the way for more efficient and adaptable solutions to complex problems, signifying a fundamental shift in the efficacy of machine learning approaches.

4.3.4 Enhancing Model Generalisation

In the ever-shifting landscape of real-world data, machine learning models face a constant challenge: maintaining adaptability. Overfitting, the enemy of generalization, lurks in every corner, waiting to trap models in the quicksand of memorized patterns. But amidst this danger, a swarm of solutions emerges - swarm algorithms. Drawing inspiration from nature's intricate dances, these algorithms empower models to strike a delicate balance between exploration and exploitation. Imagine a flock of birds, each exploring different branches of a tree, their collective knowledge weaving a tapestry of possibilities. This is the essence of swarm intelligence applied to machine learning. By simultaneously investigating diverse avenues, swarm algorithms minimize the risk of overfitting. Each individual, be it a particle in a swarm or an ant on a foraging path, contributes a piece to the puzzle, building a robust understanding that transcends specific data points. This leads to models with exceptional generalizability, capable of adapting to new situations with grace. The impact of swarm algorithms goes far beyond mere resilience. They revolutionize model optimization, opening doors to diverse applications. Imagine training a neural network not as a solitary ascent of a mountain, but as a collective exploration of a vast, interconnected landscape. This is the power of swarm-driven hyperparameter tuning, where models find their optimal configurations not through trial and error, but through the harmonious interplay of individual searches. The following sections delve deeper into the intricate workings of swarm algorithms and their specific applications in the realm of deep learning. We equip you with practical insights, enabling you to unleash the power of collective intelligence in your own machine learning endeavors.

4.4 Swarm Algorithms in Deep Learning

Deep learning, with its intricate neural networks, has achieved impressive results in a range of fields. But building and fine-tuning these networks is like navigating a vast, tangled maze. There are countless possible paths (network architectures), making it difficult to find the best one [102]. This is where swarm algorithms come in as helpful guides. Imagine a group of ants working together to find food. They explore different paths, share information, and adapt their route based on their findings. Similarly, swarm algorithms mimic the teamwork and adaptability of natural swarms to navigate the maze of possibilities in deep learning.

4.4.1 Swarm-based Neural Architecture Search

Deep learning models have revolutionized countless fields, but designing the right architecture for each task can be a complex puzzle. Luckily, a technique called Neural Architecture Search automates this process, exploring different network structures to find the one that performs best. The detailed components of the NAS is explained in Figure 4.2.

NAS typically relies on computationally expensive methods like reinforcement learning or genetic algorithms. But a powerful alternative arises from nature itself: swarm algorithms. Inspired by the collective behavior of flocks of birds and schools of fish, these algorithms efficiently navigate the vast world of potential network designs. Imagine a swarm of tiny particles flitting across a landscape, each representing a different neural network architecture. Their movements are

Figure 4.2: Components of NAS.

guided by their own past successes and the successes of their neighbors. As they explore, they gradually move towards areas with better performance, eventually converging on the most promising network design.

Finding the optimal design for neural networks can be a daunting task, involving vast combinations of layers and connections. Fortunately, swarm algorithms like PSO and ACO offer a powerful solution. These algorithms mimic the collective behavior of natural swarms, like particles or ants, to navigate this complex landscape effectively. Imagine countless tiny engineers, each exploring different designs and collaborating to find the best one [92, 257]. By guiding their search and learning from each other, swarm algorithms drastically reduce the computational burden compared to traditional methods, making it practical for real-world applications.

4.4.2 Evolutionary Algorithms in Deep Neural Network Design

Evolutionary Algorithms (EAs) have emerged as a highly promising paradigm in the design of deep neural networks (DNNs), providing a compelling alternative to conventional manual or random techniques [77]. EAs play a pivotal role in optimizing not only the architecture and hyperparameters of neural networks but also their topology, drawing inspiration from the fundamental principles of natural selection, genetic diversity, and survival of the fittest. Strategically employed in the context of DNNs, these algorithms refine and enhance neural network topologies through a subtle, iterative evolution of a population of candidate solutions. Over successive generations, EAs dynamically adapt, honing network configurations and ultimately enhancing performance in task-specific domains.

The application of EAs in DNN design involves modeling neural network topologies as individuals within a population. Subsequently, genetic operators such as crossover and mutation are applied to these individuals, generating unique configurations. Rigorous assessment of each individual's fitness is based on its performance in a specific activity, with only the most proficient individuals selectively retained to form the next generation. This iterative process continues until convergence is achieved or a predetermined stopping criterion is met. Due to the intricate nature of the task or the extensive search space encompassing numerous configurations, evolutionary algorithms prove particularly valuable in scenarios where manual design becomes impractical. The ability of EAs to effectively explore and exploit the expansive solution space constitutes a distinct advantage in their application to DNN design [10]. This capability is notably beneficial in the field of neural architecture search, where the overarching objective is the autonomous discovery of optimal neural network topologies. Compared to exhaustive search methods or arbitrary exploration, evolutionary NAS methods have demonstrated not only greater efficiency but also enhanced effectiveness. These algorithms significantly contribute to the automation of the design process by harnessing evolutionary principles, thereby alleviating the burden on human expertise and unlocking the latent potential for the creation of unique, high-performing neural network topologies.

4.4.3 Combining Swarm Algorithms and Deep Learning Frameworks

Deep learning has revolutionized various fields by automatically learning from data. However, designing high-performing models can be tricky. Finding the best structure and settings (hyperparameters) involves navigating a vast landscape of possibilities. Here's where swarm algorithms come in. Inspired by natural swarms like beehives or bird flocks, they work like teams of tiny "agents" searching for the best solution. Each agent explores, remembers what works, and shares its findings with others. Over time, they converge on the best option. The study by K. K. Buddi et al. (2023) proposed hybrid approach that used Modified Binary Dragonfly Optimization (MBFO) and Fuzzy Particle Swarm Optimization (FPSO) alongside deep learning for feature selection and reduction, ultimately achieving promising results in fake news detection [194]. Basically swarm algorithms applied to deep learning help in two key ways:

1. Finding the best model layout: Imagine a maze leading to the perfect model setup. Swarm algorithms explore countless paths simultaneously, finding the best one much faster than traditional methods.

2. Fine-tuning the settings: Think of hyperparameters as knobs controlling the deep learning engine. Swarm algorithms adjust these knobs precisely, squeezing out maximum performance [187].

This powerful combo leads to several benefits:

1. Stronger models: Swarm algorithms help build models that are more accurate and robust, making them less prone to errors.

2. Faster development: Automating hyperparameter tuning saves researchers time and effort.
3. Scalability: Swarm algorithms excel at parallel processing, making them ideal for large datasets and complex models.

This collaboration between deep learning and swarm intelligence is still young, but its potential is enormous. By harnessing the power of collective intelligence, we can build even more advanced AI systems for various applications.

4.5 Case Studies in Optimizing Deep Neural Networks

This section showcases real-world examples where swarm algorithms successfully improved deep neural networks. We'll explore applications in computer vision, image classification, and even natural language processing. These case studies highlight the practical benefits of combining swarm intelligence and deep learning, including faster convergence, more efficient training, and better accuracy. This collaboration tackles challenges like finding optimal network structures, fine-tuning parameters, and boosting overall performance. Later sections will delve deeper into practical implementations and further examples, revealing the clear advantages of this interdisciplinary approach. Combining machine learning and deep learning with swarm algorithms has proven highly effective in various practical applications. We'll now analyze specific examples from real-world scenarios, showcasing how this powerful blend of swarm intelligence and advanced learning algorithms offers exceptional flexibility and impact.

4.5.1 Case Study 1: Robotics and Automation

In Industry 4.0, where technology runs the smarter factories and industrial processes, Ocado, the online grocer, has pioneered a groundbreaking solution: swarm robotics. Their automated warehouses for online grocery delivery exemplify the seamless integration of swarm intelligence, machine learning, and deep learning. The following shows how it is works and algorithm inspired by Ocado technology is given in Algorithm 7.

1. Imagine a colossal chessboard spanning football fields, where thousands of robots move strategically. Each square holds storage bins filled with groceries. These robots collaborate harmoniously to pick customer orders, optimizing efficiency.
2. These grids, aptly named "hives," can scale up or down while maintaining precise clearances between robots. The swarm itself is scalable and one can populate the hive with bins and robots as needed. Every robot is identical, ensuring seamless task hand offs and significant economies of scale.

3. The hive dynamically reoptimizes based on future demand forecasts for each product. Machine learning algorithms adapt to external factors like weather, seasonality, and promotions. Remarkably, the hive knows nothing about groceries, opening doors to cross-sector applications [128].

Algorithm 7 Ocado-inspired Warehouse Automation with Swarm Robotics

---

Initialize:
robots: List of all robots in the warehouse
items: List of all items needing processing
tasks: Empty list of tasks
locations: Map of locations within the warehouse
while items are not processed **do**
  Generate Tasks:
  for each item $i \in items$ **do**
    Add task to *tasks* for picking and transporting i
    if item requires sorting or packing **then**
      Add task for handling i to *tasks*
    end if
  end for
  Task Allocation:
  for each robot $r \in robots$ **do**
    Use swarm intelligence algorithm (e.g., ACO, PSO) to select best task t from *tasks* for r considering:
    - Distance between r and task location
    - r's load and capacity
    - Potential for collaboration with other robots
    Remove t from *tasks* and assign to r
  end for
  Task Execution:
  for each robot $r \in robots$ with assigned task t **do**
    Execute t autonomously by r:
    - Navigate to item location
    - Pick up, transport, and manipulate item(s)
    - Communicate with other robots for coordination and obstacle avoidance
  end for
  Task Completion:
  for each robot $r \in robots$ that completed task t **do**
    Update:
    - r's status and available load
    - Item status and location
    - Remove t from *tasks*
  end for
end while

4.5.2 Drug Discovery

In the realm of drug discovery, where identifying potential compounds for treating diseases is quite the need of the hour, swarm intelligence emerges as a powerful ally. This novel approach draws inspiration from natural phenomena observed in social insects, birds, and fish. Here's how it transforms drug discovery, swarm intelligence algorithms efficiently navigate high-dimensional spaces, addressing challenges like non-convex optimization. Automatic parameter tuning and optimal architecture search become feasible. The synergy of swarm intelligence with machine learning and deep learning creates a robust hybrid model known as evolutionary machine learning. Flowchart 1 depicts the process of optimising drug discovery with swarm intelligence. Askr et al., provide an integrative review of deep learning's applications, including drug–target interactions, drug–drug similarity, and side effect predictions [12]. Bharti et al., explore swarm intelligence for deep learning, emphasizing concepts, challenges, and trends [164]. In summary, swarm intelligence bridges the gap between optimization challenges and machine learning in drug discovery. By leveraging collective behavior, we pave the way for more efficient and accurate drug development processes.

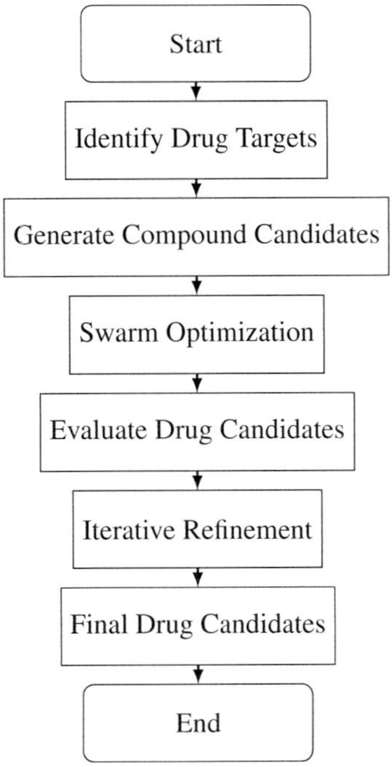

Flowchart 1. Flowchart illustrating the process of optimizing drug discovery with Swarm Intelligence.

These two examples showcase the potential of combining swarm algorithms with machine learning and deep learning. This "fusion" can not only improve network performance but also be applied to other fields like robotics, healthcare, and finance. By implementing these cutting-edge technologies, we can achieve better efficiency, optimization, and overall advancement in various sectors.

4.6 Experimental Results

The decision to put Particle Swarm Optimization in your ML or DL model depends on the tune your data plays. If the puzzle at hand is straightforward, existing algorithms like random forests or support vector machines might already hold the key [162]. Think moderate-sized datasets where understanding the model's inner workings is crucial, like in medical or regulatory applications. Here, the transparency of traditional ML models shine, allowing you to explain each step in the decision-making process. But when the data erupts in a symphony of intricacies, demanding high-dimensional analysis, it's time to invite deep learning models to the floor. These black-box beauties excel in high-data environments, automatically extracting hidden patterns through their layered architecture. But remember, their hunger for processing power and labeled data is insatiable. Ultimately, the choice to tango with PSO lies in a delicate balance. Weigh the complexity of your problem, the need for interpretability, the richness and structure of your data, and your computational resources. Only then can you decide whether traditional ML, with its transparency and simplicity, or deep learning, with its power and adaptability, will lead you to the perfect PSO match. By making use of machine learning algorithms and deep learning algorithms the following results have been found and the results are graphically represented in Figure 4.3. The integration of a basic swarm optimization algorithm that is Particle Swarm Optimization, into a machine learning algorithm (Radom Forest) using a commonly available dataset that is load digits gave the accuracy of 0.972 and the test size taken is 0.2. The integration of a basic swarm optimization algorithm with Particle Swarm Optimization, into a Deep neural network with the load digits gave the accuracy of 0.983 and the test size taken is 0.2. The hyperparameters tuned include the learning rate and the number of neurons that is Drop ratio in the hidden layer.

4.7 Challenges and Limitations

The incorporation of swarm algorithms into the realms of machine learning and deep learning, while holding immense promise, introduces a set of intricate challenges and inherent limitations. To ensure the effective application of these sophisticated methods in real-life scenarios, it is imperative to thoroughly comprehend and actively address these obstacles.The following are the common

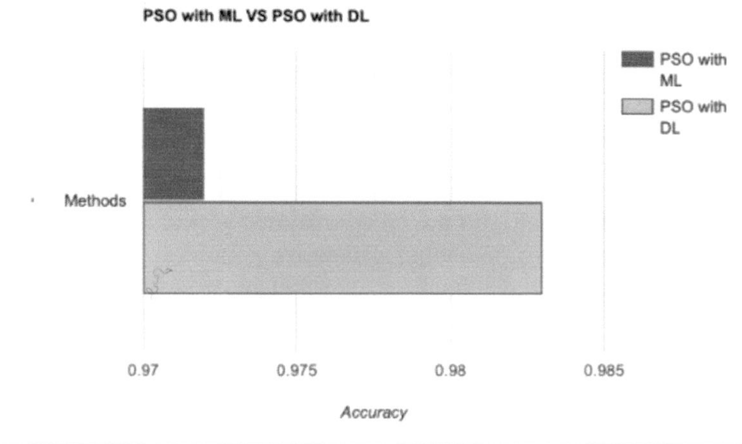

Figure 4.3: Comparison of PSO with ML and DL.

challenges and limitations while integrating swarm algorithms with machine and deep learning:

1. **Scalability Challenges:** The computational complexity of swarm algorithms occasionally exhibits quadratic or higher degrees of difficulty. Particularly when dealing with extensive datasets or intricate optimization problems, these techniques demonstrate high computational demands. As problem size increases, inherent scaling difficulties emerge, significantly impacting the efficiency and appropriateness of swarm intelligence technology.

2. **Scalability Challenges:** The computational complexity of swarm algorithms occasionally exhibits quadratic or higher degrees of difficulty. Particularly when dealing with extensive datasets or intricate optimization problems, these techniques demonstrate high computational demands. As problem size increases, inherent scaling difficulties emerge, significantly impacting the efficiency and appropriateness of swarm intelligence technology.

3. **Balance between Convergence and Exploration-Exploitation:** Premature Convergence poses a significant challenge for swarm algorithms, leading the swarm towards suboptimal solutions before reaching the global optimum. Striking the delicate balance between exploration and exploitation proves to be a complex undertaking. Inadequate parameter adjustment can result in suboptimal performance, highlighting the intricate nature of balancing these two essential components in swarm algorithm optimization.

4. **Constraints on Computational Resources and Time:** Utilizing swarm methods, particularly in conjunction with deep learning, demands substantial computational resources. The already resource-intensive process of training deep neural networks is further intensified with the incorporation of swarm intelligence. This presents challenges for scholars and professionals with limited access to high-performance computing resources.
5. **Ethical and Regulatory Considerations:** In the realm of swarm algorithms, concerns arise about categorizing these models as "black-box" entities. The inherent lack of interpretability and explainability raises ethical concerns, especially in sensitive domains like healthcare and finance. Transparency in decision-making processes is crucial, necessitating a comprehensive understanding of swarm algorithms to instill confidence and meet regulatory obligations.
6. **Hybrid Approaches and Multi-Objective Optimization: Tackling Integration Challenges:** Integrating swarm algorithms into existing machine learning workflows presents a formidable barrier. Developing hybrid approaches that successfully merge swarm algorithm attributes with alternative optimization techniques requires meticulous planning and validation. Utilizing swarm algorithms for multi-objective optimization adds complexity, demanding a careful balance amid clashing objectives.
7. **Dealing with the Difficulties of Adjusting to Changing Environments and Immediate Data:** Dynamic scenarios in practical applications pose challenges for swarm algorithms. Adapting seamlessly to fluctuations in the optimization landscape, especially with real-time data integration, requires careful consideration. In summary, while the integration of swarm algorithms with machine learning and deep learning holds promise for innovative resolutions to optimization problems, researchers and practitioners must be aware of associated impediments and limitations. Addressing these challenges is paramount, necessitating a steadfast commitment to ongoing research, continual algorithm enhancements, and a cautious approach to application design. To fully harness the potential of swarm-based techniques in the evolving field of artificial intelligence, synchronized and resolute endeavors are imperative.

4.8 Future Development Prospects and Areas for Further Investigation

The integration of machine learning and deep learning with swarm algorithms is a dynamic and continuously expanding domain that has several prospects for future investigation and progress. As researchers go further into the realm of artificial intelligence, they uncover a multitude of interesting paths and research prospects. The following provides an overview of future development:

1. **Hybrid Approaches and Multi-Objective Optimization:** Discover the intriguing realm of hybrid algorithms, which adeptly merge the unique attributes of multiple swarm algorithms or seamlessly incorporate the capabilities of swarm intelligence with conventional optimization techniques. Delve into the intricate realm of employing swarm algorithms to tackle multi-objective optimization problems, effectively resolving the difficulties of reconciling conflicting objectives in various domains.
2. **The Use of Real-Time Data in Dynamic Environments:** Enhance the functionality of swarm algorithms to efficiently function in dynamic and evolving contexts with continuous variations in data distribution and system parameters. Create adaptive swarm-based systems capable of autonomously modifying their behavior in order to enhance their versatility and usefulness in dynamic and unpredictable situations.
3. **Clarifying AI and Comprehensibility:** It is essential to acknowledge the inherent challenge of understanding swarm-based models. This involves meticulously ensuring that the decision-making processes of these algorithms are not only observable but also easily comprehensible. The primary objective is to investigate research methodologies that can effectively extract meaningful insights from swarm-based models. Our objective is to equip stakeholders with the information and assurance to have faith in and comprehend the rationale behind the results generated by these models. The primary goal is to enhance the lucidity and understandability of the decision-making procedures integrated within swarm-based algorithms.
4. **Collaboration between humans and swarms:** Investigate the possibilities of cooperative decision-making, with a particular emphasis on the integration of human experts and computers that employ swarm intelligence. Examine the intricate interplay between human intellect and the collective knowledge employed by swarm-based algorithms, uncovering the potential enhancements they could provide to decision-making procedures. The objective of this study is to understand and improve the dynamic relationship between artificial intelligence and human cognition, resulting in a cooperative and mutually beneficial partnership.
5. **Edge Computing and Swarm Intelligence:** Investigate the viability of incorporating swarm algorithms into edge computing environments to provide decentralized and efficient data processing in close proximity to its origin. Create agile swarm models that are specifically tailored for deployment on devices with limited resources, hence increasing the potential applications of swarm intelligence in edge computing.

6. **Robustness and Security:** Improve the resilience and protection of swarm-based systems against intentional attacks and interruptions. Investigate innovative strategies to enhance swarm algorithms when employed in crucial applications, guaranteeing their robustness against malevolent interference and preserving their authenticity inside vulnerable domains.
7. **Swarm Algorithms Inspired by Biological Systems:** Discover the captivating world of swarm algorithms, which are inspired by the intricate mechanics of biological systems, as you go on an exciting journey into uncharted territories. Gain significant insights by studying diverse ecosystems and their inhabitants, which may be utilized to develop cutting-edge swarm intelligence models. Harness the complex behaviors observed in nature to create sophisticated algorithms with enhanced adaptability and problem-solving capabilities. As attentive researchers investigate these potential areas, they are prepared to uncover inventive solutions to existing challenges. The integration of machine learning, deep learning, and swarm algorithms is essential for introducing a new era of intelligent systems. This interdisciplinary advancement not only has the capacity to tackle intricate real-world problems, but also propels the advancement of systems that are both intelligent and flexible. The integration of these cutting-edge technologies triggers a cascade effect, creating possibilities for innovative progress in addressing the complexities of our ever-evolving environment.

4.9 Conclusion

To conclude, the combined power of machine learning, deep learning, and swarm algorithms could boost artificial intelligence. Particle Swarm Optimisation, Ant Colony Optimisation, bee-inspired approaches, fire fly, and bat algorithms have shown novel solutions to address challenging optimisation issues. Machine learning and swarm approaches increase the creation of models and optimisation. Swarm-based methods improve machine learning model training, selection, feature engineering, and hyperparameter customisation. Collaboration speeds model convergence and improves representation durability and usefulness. The convergence of swarm algorithms and deep learning has altered automated model construction, especially Neural Architecture Search. Robotics and drug discovery case studies integrate swarm methods, machine learning, and deep learning. These applications show how versatile swarm-based methodologies are for system optimisation and tackling complicated or unsolvable practical challenges. Scalability, convergence, and ethics require careful consideration. Addressing these issues and creating ethical approaches to integrate swarm algorithms with deep learning and machine learning to succeed. Hybrid machine learning-swarm algorithm methods could be studied in dynamic, real-time environments that require adaptability. Interpretability and explainable AI must be investigated regularly to ensure model reliability utilising these integrated methodologies. In a changing environment, machine learning, deep learning, and swarm algorithms can redefine AI. We must balance technological improvements, ethical considerations, and a need to understand how AI-powered systems handle real-world complexity.

Chapter 5

Search and Rescue Using UAV SWARM in Disaster Management: A Survey

Prakarsh Srivastava and *J. Sathish Kumar**

Unmanned Aerial Vehicles (UAVs) have proven to be highly beneficial in disaster assessment, Search and Rescue (SAR), and other situational analysis in affected areas. This chapter offers an overview of the current state-of-the-art in UAV utilization for crisis-focused disaster management. It covers various types of UAVs and commonly used payloads, the different operational capabilities of these UAVs, and their applications in real-world disaster scenarios. The paper also delves into ethical scenarios and security aspects, particularly addressing potential eavesdropping by unknown UAVs. Additionally, challenges and prospects are discussed. Through a thorough examination of existing literature, the paper provides recommendations for practitioners, policymakers, and researchers on how UAVs can be effectively employed to enhance disaster response and recovery efforts.

Department of Computer Science and Engineering, MNNIT Allahabad, U.P., India.
* Corresponding author: sathish613@mnnit.ac.in

5.1 Introduction

Disasters can be natural or man-made but the potential to create drastic changes on human lives and society is severe. Loss of life, property, and other infrastructure can disrupt the social and economic fabric of affected communities. Managing such disasters is a complex and challenging task that requires effective planning, preparedness, and response efforts from a range of stakeholders including government agencies, emergency responders, and the general public. In the case of failure to manage disasters, there can be severe consequences, including long-term health and economic impacts. So, there is a critical need to improve disaster management strategies and develop innovative technologies that can help control the impact of these disasters and to grow in the field of response and recovery efforts.

One of the basic reasons for using Unmanned Aerial Vehicles (UAV) in disaster management is their ability to access hard-to-reach areas that may be unsafe or inaccessible for ground-based responders of Search and Rescue (SAR) teams [20]. UAVs can also be used to locate missing persons by providing real-time aerial surveillance of disaster zones and identifying areas where survivors may be trapped or in need of assistance. Various technologies can be equipped with UAVs to provide real-time detection of objects and other personnel [217].

UAVs drones can be classified on the basis of Landings, Aerodynamics, and Altitude. In this chapter, the focus is on the UAVs classified on the basis of aerodynamics [160].

There are several types of UAVs based on their capabilities and payloads and how they can be used in the disaster scenarios such as

- Fixed-wing UAVs: They are designed to flyover a larger range, and for longer time periods which makes these types of UAVs best for SAR operations for example in the case of an oil leak in the Oil Rig in the sea. They can be equipped with high-resolution cameras and thermal imaging sensors to provide real-time aerial surveillance of disaster zones.
- Rotary-wing UAVs: Also known as quadcopters or drones, rotary-wing UAVs are ideal for conducting close-range operations in urban areas and difficult-to-reach locations. They can also be equipped with high-definition cameras, and other sensors to provide detailed 3D maps of disaster zones and assess structural damage.
- Hybrid UAVs: This type of UAV is hybrid in nature they have the capacity to fly longer distances and can perform close-range operations which are the properties of the two above UAVs type. They can be equipped with payloads such as medical supplies, communication devices, and emergency supplies to assist disaster victims.

- Tethered UAVs: These UAVs are attached to a stationary ground-based platform by a cable, allowing them to hover over a specific location for extended periods of time. They can be used for aerial surveillance, and communication, and to provide real-time data feeds to disaster response teams.

While UAVs can be used in many scenarios their usage must be guided for ethical considerations. In the case of Swarms of UAVs there can be a chance of unknown UAVs becoming part of the swarm and corrupting the network of swarms.

UAV operators must be trained and licensed to operate the vehicles, and the vehicles themselves must meet certain safety standards. UAVs can provide valuable data and insights, but their use must be guided by appropriate ethical principles. For example, UAV operators must ensure that they do not violate the privacy rights of disaster victims or inadvertently cause harm by flying over sensitive areas.

The purpose of this survey paper is to provide a comprehensive overview of the usage of Unmanned Aerial Vehicles (UAVs) for disaster management. The scope of the paper will cover the different types of UAVs used in disaster management, their capabilities and payloads, and the challenges that arise in their deployment. The paper will also address the regulatory and ethical considerations that must be taken into account when using UAVs for disaster management. This paper also includes the Dynamic flow analysis during a flood scenario. This paper also tries to imply the dynamic routing of UAVs designed to efficiently cover the affected area as post-disaster management. In Section (2) Types of UAVs and their capacities are discussed along with the ethical considerations. Section (3) comprises UAV operations and a proposed method for an efficient and wider coverage with energy saving.

5.2 UAV Components and Applications

In disaster management, UAVs utilize a range of components that enable their effective operation and data collection capabilities. These components include cameras, sensors communication systems, and navigation equipment. Cameras, such as RGB (Red, Green, Blue) and thermal imaging cameras, are extensively used for aerial surveillance, damage assessment, and search and rescue operations. Sensors, such as LiDAR (Light Detection and Ranging), provide detailed 3D mapping of disaster-affected areas. Communication systems ensure seamless data transmission between the UAV and ground stations, enabling real-time monitoring and control [123]. Navigation equipment, including GPS (Global Positioning System) receivers and inertial measurement units, facilitate accurate positioning and flight control. These components work synergistically to enhance the capabilities of UAVs in disaster management, enabling rapid and comprehensive data collection, analysis, and decision-making.

5.2.1 UAV Components and Ethical Considerations

These operations must be done taking ethical considerations. There may be a chance of an eavesdropper drone in the swarm of legitimate drones. They might hinder the privacy and safety of the data [121]. In the consideration of real time ethical scenarios and security aspects there can be various cases such as

- **Informed Consent:** When surveying a disaster prone area ethical consideration in such cases is obtaining informed consent from residents before conducting aerial surveys that may capture personal property or vulnerable situations.
- **Equitable Deployment:** UAVs are primarily used in urban areas, neglecting rural or remote locations. Ensuring equitable distribution of UAV resources to avoid leaving certain populations underserved during disaster response is the ethical consideration in such scenario.
- **Data Encryption:** Transmission of real-time data to ground control during disaster operations needs consideration of robust encryption methods to secure communication channels and prevent unauthorized access to sensitive information. Digital Image Watermarking is a prominent solution for sensitive data transmission over UAV communication network. Taking into the consideration that the encryption and the decryption does not take much computational power [1].
- **Public Perception:** Drones should be deployed after clearly explaining their purpose in search and rescue, damage assessment, and supply delivery helps alleviate fears and ensures community cooperation. This transparency respects individuals' rights and builds trust between disaster response teams and the local population.

There are several methods to ensure such illegal activity (drones can be preprogrammed with random noise which is only known to legitimate drones this can be used to identify the unknown UAVs).

There are several ethical considerations that should be taken care of while using UAVs in disaster management:

(1) Privacy (2) Data Security (3) Safety and Liability (4) Consent and Public Perception (5) Equity and Access (6) Regulatory Compliance.

Adequate training and certification of UAV operators should be in place to ensure compliance and responsible use of the technology. It is crucial to establish clear guidelines and protocols for data collection to ensure the respect of individual privacy. Considerations should be given to the handling and storage of collected data, as well as obtaining consent when necessary, issues of liability need to be addressed, determining who is responsible in case of accidents or damages caused by UAVs [48]. Obtaining consent from individuals and communities for UAV operations, especially in sensitive areas or during search and rescue operations, is crucial [69]. When transmitting sensitive information using drone channels, such as images including the sensitive social media contents watermarking the image during transmission ensures the security and confidentiality. Use of Redundant Discrete Wavelet Transform (RDWT) and Singular Value Decomposition (SVD) for watermarking in less detailed areas, encrypts with a pseudo-random key, and optimizes scaling for balance [2].

There are different types of UAVs commonly used in disaster management, including fixed-wing, rotary-wing, and hybrid UAVs. Each type of UAV has its own advantages and limitations, and its suitability for different disaster scenarios varies based on various factors such as terrain, weather, and the nature of the disaster.

There are various components of UAVs Cameras, Multisensor Payloads, GPS receivers and trackers, Communication Systems, On-Board Processing Units, Telemetry Systems, and Payload Delivery Systems. Table 5.1 below contains the operation-based components description of the components used in drones [32, 8].

Table 5.1: Components used in UAVs for Disaster Management.

Operation	Components	How They Contribute
	High-resolution cameras	Capture images and videos for assessing damage
Damage Assessment	LiDAR sensors	Generate detailed 3D maps of affected areas
	Thermal imaging cameras	Identify hotspots or areas of potential danger
	GPS receivers	Geotag collected data for accurate mapping
	RGB cameras	Provide live video feeds for locating survivors
Search and Rescue	Thermal imaging cameras	Detect body heat signatures for identifying individuals
	Multi-sensor payloads	Combine different sensors for enhanced search capabilities
	GPS trackers	Track the location of the drone and potential victims
	Multi-sensor payloads	Gather real-time data on weather conditions, air quality, etc.
Situational Awareness	Communication systems	Establish wireless communication links in remote areas
	Onboard data processing	Analyze collected data for situational understanding
	Telemetry systems	Transmit vital information to ground control

5.2.2 *Application of UAV in Disaster Management*

- **Fixed-Wing UAVs:**
  - 1999: The US Federal Emergency Management Agency (FEMA) used a fixed-wing UAV called the AeroVironment Pointer to survey damage in the aftermath of Hurricane Floyd.
  - 2004: The European Commission's Joint Research Centre deployed a fixed-wing UAV called the UAV-Eye in the aftermath of the Indian Ocean tsunami. The UAV was used to survey the damage to infrastructure and buildings in affected areas.
  - 2010: The US National Oceanic and Atmospheric Administration (NOAA) used a fixed-wing UAV called the Aerosonde to gather data on Hurricane Earl. The UAV was equipped with sensors that measured wind speed, temperature, and humidity, and provided valuable data on the storm's intensity and trajectory.
  - 2011: Following the earthquake and tsunami in Japan, the Japanese government deployed a fixed-wing UAV called the RQ-4 Global Hawk to survey the damage and assess the impact on infrastructure and the environment.
  - 2013: After the Uttarakhand floods, the Indian Air Force used Heron and Searcher II fixed-wing UAVs to help locate survivors and assess the damage caused by the disaster. (They also used Rotary Wing UAVs).
  - In 2014, after a landslide in Pune, Maharashtra, the Indian Army used a Searcher II fixed-wing UAV to help locate survivors and assess the damage caused by the disaster.
  - 2015: The World Food Programme used a fixed-wing UAV called the SenseFly eBee to survey damage in the aftermath of the earthquake in Nepal. The UAV was used to create detailed maps of affected areas, which were used to identify areas of need and coordinate relief efforts.
  - 2017: The US Federal Aviation Administration (FAA) granted a special exemption to a company called Zipline to operate fixed-wing UAVs in disaster zones in the US. The UAVs were used to deliver medical supplies and other essential items to areas affected by natural disasters.
  - In 2017, after heavy rain and flooding in Mumbai, the Mumbai Police used a Panchi fixed-wing UAV to monitor the situation and provide real-time updates to the authorities.

- 2018: The Indonesian government used fixed-wing UAVs to survey the aftermath of a 7.5 magnitude earthquake and tsunami in the city of Palu, allowing them to assess the extent of the damage and prioritize rescue efforts.
- In 2018, following the Camp Fire in California, the California Department of Forestry and Fire Protection used fixed-wing UAVs to assess the damage and identify areas that required further attention.
- 2019: Fixed-wing UAVs were used by the Australian government to map and assess the damage caused by bushfires across the country.
- In 2019, the World Food Programme used fixed-wing UAVs to assess the damage caused by Cyclone Idai in Mozambique and to map out the affected areas to help with relief efforts.
- In 2019, after Cyclone Fani hit Odisha, the National Disaster Response Force (NDRF) used a Heron UAV to survey the affected areas and assess the damage caused by the disaster.
- 2020: The Indian Air Force used a fixed-wing UAV to survey the damage caused by floods in the state of Assam, and to identify areas where relief efforts were needed.
- In 2020, during the COVID-19 pandemic, the Indian government used fixed-wing UAVs to deliver medical supplies and other essential goods to remote areas and to enforce lockdowns in certain areas.
- 2021: After Cyclone Tauktae hit the western coast of India, the Indian Navy deployed Panchi fixed-wing UAVs to survey the affected areas and assess the damage caused by the disaster.

- **Rotary-Wing UAVs:**
  - 2005: The US Navy deployed a rotary wing UAV called the Fire Scout to support search and rescue efforts in the aftermath of Hurricane Katrina.
  - 2008: The Indian government used rotary wing UAVs to assess damage and provide relief in the aftermath of Cyclone Nargis in Myanmar.
  - 2010: The UN used rotary wing UAVs to assist with relief efforts in Haiti following the devastating earthquake.
  - 2013: Rotary-wing UAVs were used to survey damage caused by Typhoon Haiyan in the Philippines.

- In 2013, during the Uttarakhand flash floods, the Indian Air Force deployed rotary wing UAVs to conduct search and rescue operations.
- 2015: Rotary-wing UAVs were used to survey damage caused by the Nepal earthquake. They were proven to be very effective. The UAVs were able to provide high-resolution imagery and identify areas where rescue efforts were needed.
- In 2015, after the earthquake in Nepal, India deployed rotary wing UAVs to assess the damage and provide relief to affected areas.
- 2016: Rotary-wing UAVs were used in the aftermath of the Ecuador earthquake to help locate survivors and assess the damage.
- 2017: Rotary-wing UAVs were used in the response to Hurricane Harvey in the United States, primarily for search and rescue and damage assessment. The US Federal Aviation Administration (FAA) authorized the use of rotary-wing UAVs to aid search and rescue operations.
- 2018: Rotary-wing UAVs were used in the response to the Sulawesi earthquake and tsunami in Indonesia for damage assessment and search and rescue operations.
- In 2018, after the floods in Kerala, India used rotary wing UAVs to conduct aerial surveys and assess the extent of the damage.
- 2019: Rotary-wing UAVs were used in the response to Cyclone Idai in Mozambique for damage assessment, search and rescue, and delivery of supplies.
- In 2019, during Cyclone Fani, Puri, Odisha. India deployed rotary wing UAVs to provide real-time information on the extent of the damage caused by the cyclone.
- 2020: Rotary-wing UAVs were used in the response to the Australian bushfires for mapping and monitoring hotspots and identifying areas for evacuation.
- In 2020, after the gas leak in Visakhapatnam, India used rotary wing UAVs to assess the damage and conduct relief operations.
- In 2020, rotary-wing UAVs were used by the Indian Army to survey the aftermath of a glacier burst in the northern state of Uttarakhand, allowing them to locate and rescue trapped survivors.

- 2021: During the Uttarakhand flash floods, India deployed rotary wing UAVs to provide real-time information on the extent of the damage and assist in search and rescue operations.

- **Hybrid UAVs:**
  - 2011: Hybrid UAV was used to map the extent of damage to the Fukushima nuclear power plant. The UAV was able to cover a large area and provide high-resolution imagery that helped in the cleanup and recovery efforts [52].
  - 2015: Hybrid UAVs have been used in earthquake response efforts in Nepal.
  - 2019: A hybrid UAV was used by the South African National Defence Force (SANDF) to assist in disaster response efforts following heavy floods in the Eastern Cape region of South Africa. The hybrid UAV was used to survey the affected areas and create detailed 3D maps that were used to aid rescue and relief operations.
  - In 2019, the Indian Army used a hybrid UAV developed by ideaForge to assess flood-damaged areas in Karnataka. The UAV was able to capture high-resolution images and create 3D maps of the affected areas.
  - 2020: A hybrid UAV was used by the Indian Navy to assist in disaster response efforts following the devastating floods that hit the state of Kerala in India. The hybrid UAV was used to conduct search and rescue operations in flood-affected areas and provide real-time situational awareness to the rescue teams on the ground.
  - In 2020, the Indian Navy used a hybrid UAV developed by the Indian Institute of Technology (IIT) Hyderabad to aid in disaster response efforts following Cyclone Amphan. The UAV was equipped with a thermal camera and was used to assess the extent of damage in affected areas.
  - 2021: A hybrid UAV was used by the United Nations Office for the Coordination of Humanitarian Affairs (UNOCHA) to assist in disaster response efforts following the eruption of the Mount Nyiragongo volcano in the Democratic Republic of Congo. The hybrid UAV was used to survey the affected areas and assess the damage caused by the eruption.
  - In 2021, the National Disaster Response Force (NDRF) used a hybrid UAV developed by ideaForge to survey flood-affected areas in Maharashtra. The UAV was able to capture high-resolution images

and create 3D maps of the affected areas, aiding in rescue and relief efforts.

5.3 Operations and Algorithms

5.3.1 Operations

There are many operations that can be performed using a UAV or a swarm of UAVs, but in the field of disaster management [223]. By implementing efficient algorithms for image processing, pattern recognition, and data fusion, drones can provide real-time information to decision-makers, enabling faster and more informed response strategies, keeping such accords here are various types of operations that can be performed:

- **Damage Assessment and Mapping:**
  With the use of cameras and sensors, the detection of the affected area can be done easily and swiftly [29]. They capture aerial imagery and collect data to create detailed maps, that provide the needed information for immediate SAR operations [54]. These maps help identify affected areas, prioritize resource allocation, and aid in developing response plans [71].
- **Search and Rescue Operations (SAR):**
  Drones are mostly used in SAR operations, particularly in hazardous or inaccessible environments. Cameras and thermal imaging sensors can provide aerial assistance to locate survivors and missing persons. UAVs cover large areas quickly, improving response times and increasing the chances of locating individuals in distress. They can also deliver essential supplies, such as medical kits or food, to remote or hard-to-reach locations.
- **Situational Awareness and Emergency Response:**
  The dynamic nature of disaster must be analyzed in real-time. Such as in case of a flood the road map situation of the affected area must be telecasted to the rescue vehicles carrying supplies so they can be updated for the dynamic routing accordingly [126]. UAVs equipped with cameras, video feeds, and sensors can monitor disaster events such as fires, floods, or chemical spills. By providing aerial views and data on the evolving situation, UAVs enable emergency responders to make informed decisions, allocate resources effectively, and coordinate response efforts.
- **Communication and Connectivity:**
  Disasters often disrupt or damage communication infrastructure, which becomes an obstruction for the rescue team. UAVs can act as communication relays, establishing temporary communication networks in disaster-

stricken areas. Equipped with communication devices, they provide connectivity and facilitate information exchange among response teams, enhancing coordination and ensuring effective communication during critical operations. Setting up the base station and telecasting the data from drones to the rescuers is done first.

- **Hazardous Material Management:**
  UAVs equipped with specialized sensors can monitor and assess hazardous material spills or leaks during disasters. By collecting data on the type, concentration, and spread of hazardous substances, they assist in managing and mitigating the associated risks. By analyzing these data we can work on material response, ensuring the safety of response personnel and minimizing environmental damage.

- **Damage Assessment for Infrastructure:**
  This is like a specialized task of damage assessment for infrastructure. UAVs can be used to inspect and assess the condition of critical infrastructure such as bridges, buildings, and roads after a disaster. They provide a safer alternative to manual inspections, especially in situations where physical access is restricted or dangerous.

- **Environmental Monitoring:**
  UAVs equipped with specialized sensors can monitor and assess environmental factors during and after disasters. With the help of these advanced sensors, they can collect data on air quality, water contamination, or natural resource impacts, providing valuable insights into the environmental effects of the disaster. This data helps authorities and environmental agencies make informed decisions regarding environmental protection and restoration efforts.

5.3.2 Coverage Methods

Coverage of the area is dependent on factors like camera type, altitude of drones, terrain of the area and path of the flight. Taking these factors into consideration various methods can be adapted to cover a specific area. The following figures show different types of coverage on the MNNIT Allahabad Athletic Ground. Figure 5.1 shows the grid coverage, camera type is Flir Vue 640 19 mm. Figure 5.2 shows the Cross Grid Coverage of a irregular polygon shaped area, path is shown by arrows and waypoints. Figure 5.3 shows the spline design that is used to cover complex curves and surfaces, only preliminary curves are shown here.

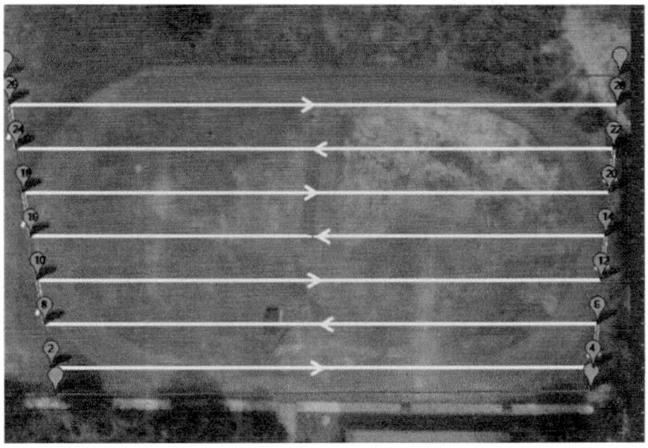

Figure 5.1: Grid coverage.

Figure 5.2: Cross grid coverage.

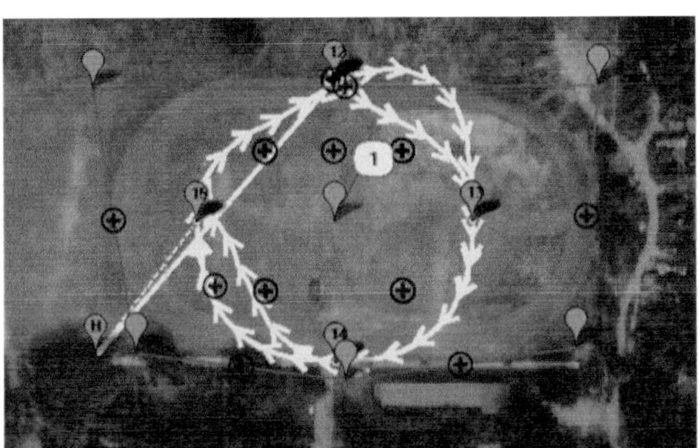

Figure 5.3: Spiral coverage.

The following are a few Coverage methods:

- **Grid-Based Coverage:**
  In this algorithm, the area to be covered is divided into a grid, and the drone follows a predefined path to systematically cover each grid cell. This can be achieved by traversing the grid row by row or column by column. This approach ensures complete coverage of the area but may result in redundant coverage in some areas. Alternatively, more advanced algorithms like the Traveling Salesman Problem (TSP) can be utilized to optimize the path.

- **Spiral Coverage:**
  The spiral coverage algorithm involves the drone flying in a spiral pattern starting from the outer boundary of the area to be covered and gradually moving inward. The drone determines the center point of the area and gradually decreases the radius of the spiral to cover the entire area. The angle of rotation can be adjusted to achieve the desired overlap and density of coverage. In case of an area that is irregular in shape, adjustment in the movement of the drone will be needed at run time.

- **Voronoi Diagram-Based Coverage:**
  The Voronoi diagram-based coverage algorithm divides the area into cells, with each cell assigned to a specific drone. The boundaries of the cells are determined based on the proximity to the drones' initial positions. Each drone covers its assigned cell, ensuring that the entire area is covered without overlapping. This algorithm is suitable for multi-drone deployments and provides efficient coverage. The complexity of computing the Voronoi diagram increases with the number of drones and irregularly shaped areas.

- **Sweep Coverage:**
  In the sweep coverage algorithm, the drone flies in parallel lines, back and forth, over the area to be covered. It moves from one end of the area to the other, covering the entire width in each pass. This algorithm is particularly useful for long, narrow areas and provides systematic coverage. This algorithm might suffer in the case of an area with more objects as they travel in a straight line or in the case when the shape of the coverage area is irregular.

- **Randomized Coverage:**
  In randomized coverage, the drone moves in a random or semi-random manner throughout the area, ensuring coverage of different regions. This approach can be useful when rapid coverage is required or when the area has complex terrain or obstacles that may hinder systematic coverage. It follows probabilistic distribution patterns to cover different regions. It

overcomes the problem of the sweep coverage algorithm but the lack of a systematic coverage method there might be a chance that some areas may remain uncovered and there is a challenge in tracking the coverage too.

- **Adaptive Coverage:**
  This is the most optimized algorithm so far. Adaptive coverage algorithms dynamically adjust the flight path of the drone based on real-time information and feedback. They optimize the coverage path based on environmental factors, sensor data, or specific objectives. These algorithms can adapt to changing conditions, prioritize certain areas, and optimize the use of resources.

In practice, a combination of different algorithms or a hybrid method might be useful in different scenarios of disaster management. Some papers discuss the potential benefits and challenges of using drones for communication purposes, such as extending network coverage to remote areas, providing temporary network infrastructure for disaster-stricken regions, and enabling communication in areas with limited terrestrial infrastructure.

In the field of coverage area, suppose the ground control station establishing is not possible like in the case of oil mining in the ocean if any anomaly occurs establishment of a ground control station is not possible drones, rather powerful ones may have to travel miles in order to reach the affected area and supply the information to the rescue team, this means that the drone needs a long battery life, and network coverage might need a satellite network. And the communication to the SAR team might need radio wave frequency which is highly vulnerable.SAGINs space air-ground networks are heterogeneously used in the case of different geological locations of the affected area [262, 103].

5.3.3 Routing Algorithms for SWARM

Routing protocols and algorithms are needed in drone swarm applications to manage the movement and coordination of multiple drones in a systematic manner [192]. These algorithms help optimize the flight paths of the drones, distribute tasks efficiently, and ensure that each drone operates within its capacity. By reducing the energy consumption of each drone, routing algorithms can also help extend the battery life and increase the overall flight time of the swarm [139]. This makes it possible for drones to cover larger areas and perform more complex tasks, which is especially useful in disaster management scenarios. Algorithms can be developed to adaptively optimize drone-to-drone and drone-to-ground communication networks in dynamic disaster scenarios. Including algorithms that dynamically adjust communication paths, select appropriate frequencies, and prioritize critical communication links to maintain reliable connectivity.

Dynamically adjusting flight routes and task assignments to optimize resource utilization, minimize redundancy, and maximize overall operational efficiency. Dynamically adjusting flight parameters, such as altitude and speed, based on real-time conditions, to maximize flight time and coverage. Additionally, algorithms can suggest optimal recharging or battery swap locations to minimize downtime. In navigation, there are some optimizations that can be done for better movement of drones while a coverage task.

Autonomous navigation and obstacle avoidance: Advanced algorithms can enable drones to navigate autonomously and avoid obstacles in real time. By incorporating machine learning and computer vision techniques, drones can detect and avoid objects, buildings, and other hazards, ensuring safe and efficient flight operations in disaster-affected areas.

The following are some benefits of using these algorithms:

Redundancy and fault tolerance: If one drone in the swarm fails or encounters an obstacle, others can take over its tasks. In case of storms and earthquakes and other infrastructure damage like disasters close-range tasks may require such dynamic scheduling of responsibilities. The redundancy provided by the swarm ensures that the mission can still be accomplished even if individual drones face issues.

Enhanced coverage and efficiency: Swarm drones can cover larger areas and complete tasks more efficiently compared to individual drones. Drone coverage can be done by many methods, this is a prime area to work on for enhanced coverage. They can work in parallel, divide complex tasks into smaller sub-tasks, and accomplish them collectively, resulting in faster response times and improved overall performance.

Scalability: Swarm drone systems can scale up or down depending on the requirements of the disaster scenario. They can easily accommodate additional drones or adjust the swarm size to match the needs of the mission. Scaling the drone capabilities can help in many cases such as in ocean surveys, and in the remote stations where the reachability of the ground station is not much probable.

Adaptability and resilience: Swarm drones exhibit adaptive behavior, allowing them to respond dynamically to changing conditions. They can reconfigure themselves, adjust their routes, and redistribute tasks in real time, making them more resilient and effective in dynamic and uncertain disaster environments. In case of flood many simulations are available that can be used to train a drone, a virtual environment is built on which drones are tested. These drones are then deployed for the real world, and dynamically changing disasters like floods force the rerouting of the search and rescue team path, this knowledge can be telemetered from the affected area to the rescue team based on the intel of drones [261, 82].

Table 5.2: Comparison of routing algorithms in Swarm UAVs.

Routing Algorithm	Description	Time Complexity	Typical Usage Scenario
Centralized Routing	Central controller manages routing for the entire swarm	$O(n)$	Large-scale swarm operations with a central command
Decentralized Routing	Individual drones make local routing decisions	$O(1)$	Swarm operations with distributed decision-making
Distributed Routing	Drones share information and collaboratively make decisions	$O(n)$	Cooperative swarm operations
Dynamic Reconfiguration	Routing plans adapt to changes in swarm configuration	$O(n)$	Scenarios with frequent addition/removal of drones
Ant Colony Optimization	Inspired by foraging behavior of ants, uses pheromone trails	$O(n^2)$	Optimizing route selection in complex environments
Genetic Algorithms	Evolves and optimizes routes using genetic operators	$O(g \cdot n \cdot m)$, g = number of generations	Route optimization and task assignment
Dynamic Task Assignment	Real-time task assignment based on drone capabilities	$O(n)$	Adaptive task allocation in dynamic environments
Cooperative Path Planning	Drones collaboratively plan routes for efficient coordination	$O(n^2)$	Collaborative operations and coverage optimization

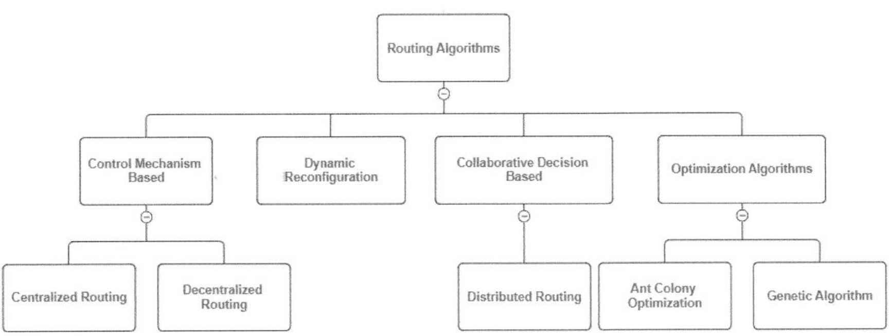

Figure 5.4: Routing algorithm categorization.

Collaborative Data Analysis: Swarm drones can collaborate in data analysis tasks by pooling their sensor data and processing capabilities. This enables a more accurate and comprehensive analysis of the collected information, leading to better decision-making and situational awareness.

Here is a brief overview of these routing algorithms

- **Centralized Routing:** A Central controller sees the entire swarm for example, Air traffic control systems guiding drones at an airport. A central controller manages the takeoff, landing, and routing of each drone to optimize overall air traffic.

- **Decentralized Routing:** Each drone in a swarm is free to make local decision for independent routing such as, delivery drones navigating a city independently. Each drone decides its route based on local information, like traffic conditions and obstacles, without relying on a central command.
- **Distributed Routing:** Coordinated Drone movement occurs when each drone in a swarm shares the information about obstacles in the path ahead for optimal path, for example, a swarm of drones used in agriculture for crop monitoring. Each drone collects data and collaboratively shares information about crop health, allowing the swarm to collectively optimize its coverage.
- **Dynamic Reconfiguration:** Search and rescue drones adapting to changes in a disaster-stricken area. Drones dynamically modify their routes based on evolving conditions to maximize the efficiency of the rescue operation.
- **Ant Colony Optimization:** As name suggests inspired by the ant searching for food. Surveillance drones patrol an area for security, employing a strategy where they create virtual "pheromone" trails. Other drones then follow these trails, contributing to the collective optimization of surveillance routes over time.
- **Genetic Algorithms:** Drones explore different routes (genetic variation), and successful routes are combined (crossover) or mutated to create new optimized routes for efficient navigation. In package delivery these genetic algorithms can be used.
- **Dynamic Task Assignment:** Disaster response drones in a dynamic environment. Drones dynamically take on tasks such as search, rescue, or medical aid based on real-time demands and their specific capabilities. It is like a team of workers dynamically adapting roles based on their skills and the evolving needs of a project.
- **Cooperative Path Planning:** Environmental monitoring drones in a forest. Drones collaborate to plan routes, ensuring efficient coverage for data collection while avoiding collisions and optimizing battery usage.

5.3.4 *Drone Simulators and UAV Models*

AirSim, Gazebo, UAVSim, PX4 SITL, Velodyne VeloView, VirtualRobotX (VRX) Simulator, Webots and ArduPilot SITL

These are some simulators that can be used for simulating a life like scenario of a disaster and the model can be trained and tested over it before performing in the actual world. AirSim supports various platforms, including DJI drones, and offers features like sensor simulation (RGB cameras, depth cameras, LiDAR, etc.), physics-based simulation, and customizable environments. It has extensive APIs and integration with popular machine learning frameworks like TensorFlow and PyTorch. Gazebo supports the integration of various drone models and provides plugins for simulating sensors such as cameras, LiDAR, and GPS. UAVSim is a lightweight, web-based simulator for drones that focuses on simplicity and ease of use [208, 250]. PX4 SITL supports various drone configurations, sensor simulations, and realistic physics-based flight dynamics. VeloView focuses on LIDAR simulation. VeloView offers advanced features for creating and customizing LiDAR scenarios, making it suitable for testing and training LiDAR-based drone applications in disaster management. It offers a wide range of sensors, physics-based simulation, and an extensive library of 3D models for creating virtual disaster management environments. Mission Planner is a ground control station software commonly used with ArduPilot. While primarily a mission planning and monitoring tool, Mission Planner also includes simulation capabilities. It allows users to simulate drone flights, test mission plans, and evaluate the behavior of ArduPilot-controlled drones in a virtual environment.

The following is the tabular representation of UAV Models that are used in the disaster management.

Table 5.3: UAV models and capabilities in Disaster Management.

UAV Model	Capabilities
DJI Phantom Series	Aerial photography, damage assessment, search and rescue support, real-time situational awareness
DJI Matrice Series	Aerial mapping, damage assessment, thermal imaging, search and rescue, payload delivery
SenseFly eBee X	Aerial mapping, large area coverage, vegetation assessment, damage assessment
Lockheed Martin Indago	Situational awareness, search and rescue, surveillance, damage assessment
Parrot Bebop 2 Power	Aerial photography, real-time video streaming, situational awareness
Yuneec Typhoon H Pro	Aerial photography, damage assessment, search and rescue, real-time video transmission

5.4 Challenges and Future Scope

5.4.1 Challenges

Despite having many advantages of drones in disaster management there are still some limitations that are limiting the current deployment of drones in various environments. Such as the limitation of Lithium-ion batteries [70].

Existing regulations often restrict the use of drones, especially in disaster-stricken areas, due to concerns over airspace safety and privacy. Developing clear guidelines and regulatory frameworks specific to disaster scenarios is crucial for wider adoption.

Many drones have limited battery life and range, which constrains their ability to cover large areas or remain operational for extended periods. Advancements in battery technology and autonomous charging stations can help overcome these limitations.

Strong winds, heavy rain, and other adverse weather conditions can impact drone operations. Improving the durability and stability of drones to operate in harsh environments is necessary for effective disaster response.

Drones currently have limited payload capacity, which restricts their ability to transport heavy or bulky supplies. Developing drones with higher payload capabilities can improve their effectiveness in delivering critical resources.

Gathering vast amounts of data through drones is only valuable if there are effective systems in place to process, analyze, and interpret the information. Developing automated data analysis tools and integrating drones into existing disaster management systems is crucial.

When taking communication based challenges into consideration the medical field is the most discussed one. In the medical field, doctors are communicating and performing remote diagnosis, thus the records of patients are vulnerable to tampering. Encryption of such sensitive data during transmission is one of the research areas. Reversible watermarking is used to ensure quick embedding and extraction times for securing medical information [3].

Although various optimization can be used to handle such problems. Software-based development in algorithms such as defining the optimized way of covering the area, and in case of a swarm of drones switching the responsibilities from one drone to another on some threshold battery levels can help a lot. Some challenges are discussed here.

- **Technical Challenges in Drone Operations:**

  Battery life and flight endurance: Increasing the flight time and endurance of drones to cover larger areas and perform extended missions. Payload capacity and flexibility: Developing drones capable of carrying heavier payloads for advanced sensors, communication equipment, or medical supplies. Autonomy and obstacle avoidance: Enhancing drone autonomy to navigate complex environments, avoid obstacles, and adapt to dynamic conditions. Communication and data transmission: Improving communication systems for reliable and real-time data transmission between drones, ground stations, and disaster management centers.

- **Integration of Drones with Existing Disaster Management Systems:**

  Standardization and interoperability: Establishing common protocols, interfaces, and standards for seamless integration of drones into existing disaster management systems. Data integration and analysis: Developing methods to effectively integrate and analyze data captured by drones with other sources, such as satellite imagery or ground-based sensors. Collaboration and coordination: Establishing frameworks for collaboration and coordination among drone operators, emergency responders, and other stakeholders to optimize drone deployments during disasters.

- **Opportunities for Improvement and Future Research Directions:**

  Improved sensor technologies: Advancing sensor technologies for enhanced imaging, thermal detection, gas detection, and other specialized capabilities to improve situational awareness and damage assessment. Artificial intelligence and machine learning: Utilizing AI and ML algorithms to automate image analysis, object detection, and decision-making processes in real time. Swarm intelligence: Exploring the potential of drone swarms for collaborative operations, increased area coverage, and coordinated response in disaster management. Human-Drone Interaction: Investigating human factors, user interfaces, and operator training to enhance the usability, trust, and acceptance of drones in disaster management scenarios. Regulatory frameworks and policies: Developing comprehensive regulations and policies that balance safety, privacy, and operational requirements to support the integration and widespread adoption of drones in disaster management [209].

5.4.2 Future Scope

These are some fields where the UAV's functionality and usage can be explored and the scope of developing new technologies and algorithms for disaster management applications.

Rapid and Remote Assessment: Drones can be used to provide a fast report of the affected area, allowing for a timely and targeted response. Future developments may focus on enhancing the capabilities of drones to capture high-resolution images, conduct real-time video streaming, and utilize advanced sensors for improved data collection and analysis [241].

Autonomous Operations: Advancements in drone autonomy will enable them to perform more complex tasks with minimal human intervention. Future drones may incorporate advanced artificial intelligence and machine learning algorithms to autonomously navigate challenging terrains, detect and avoid obstacles, and adapt to changing environmental conditions [124].

Swarm Technology: Swarm drones, working collaboratively, can cover larger areas, improve search and rescue operations, and enhance data collection efficiency. Future research may focus on developing swarm intelligence algorithms to enable coordinated actions, information sharing, and distributed decision-making among a group of drones [184].

Enhanced Sensor Capabilities: The addition of advanced sensors to the UAVs in the future will be beneficial, but the problem is managing the power consumption in these sensors. Future drones may incorporate advanced imaging technologies, hyperspectral sensors, gas detectors, and environmental monitoring instruments to provide comprehensive and real-time information about disaster-affected areas.

Improved Communication and Connectivity: Communication between drones is an important aspect and the reliability and security of data transfer between them is a matter of concern [56]. Future developments may focus on enhancing communication capabilities, such as long-range communication, secure data transmission, and integration with existing communication infrastructure, to enable seamless coordination and information exchange [185].

Data Integration and Analytics: UAVs can be used to collect a large quantity of data so in the future this large quantity of data can be customized, efficiently used, and transformed into actionable insights. Advancements in data integration and analytics will enable more efficient processing, interpretation, and visualization of drone-captured data. Integration with Geographic Information Systems (GIS) and artificial intelligence algorithms will provide valuable insights for decision-making and resource allocation.

Collaborative Integration with Existing Systems: Future developments will focus on seamlessly integrating drones into existing disaster management systems and workflows. This integration will involve developing interoperability standards, establishing protocols for data sharing and coordination among drones and ground-based assets, and ensuring compatibility with established emergency response frameworks [31].

Policy and Regulatory Frameworks: The rising needs of UAVs will be causing the need for robust policies and regulations governing their deployment. Future developments will involve the establishment of clear guidelines for drone operations, airspace management, privacy protection, and safety standards to ensure responsible and effective drone use in disaster management scenarios [224].

5.5 Conclusion

In conclusion, the utilization of drones in disaster management has revolutionized response and recovery efforts. Drones are very capable of performing the task in the long and short range and in a variety of scenarios such as in case of disaster management pre, during, and post-floods, earthquakes, wildfires, and other natural calamities. Drones offer a range of capabilities that enhance situational awareness, aid in search and rescue operations, enable rapid damage assessment, and facilitate the delivery of essential supplies in disaster-stricken areas. In this survey paper, we explored the key aspects of drone usage in disaster management, including the components used in drones, types of operations performed, ethical considerations, and the role of simulation-based training. The challenges faced in integrating drones into existing disaster management systems and identified opportunities for improvement and future research directions are discussed. Routing algorithms in swarm UAVs play a critical role in optimizing communication, resource utilization, and battery life management. Some Coverage algorithms are also discussed in this research paper. These algorithms determine efficient flight paths, distribute tasks among drones, and dynamically adapt to changing conditions. By maximizing energy efficiency and ensuring effective coordination within the swarm, routing algorithms contribute to extended flight times and enhanced operational capabilities in disaster scenarios.

While drones have demonstrated their potential in disaster management, there are still areas for further development. Regulatory frameworks need to be refined to address safety, privacy, and airspace management concerns. Technological advancements should focus on increasing flight endurance, payload capacity, and autonomous capabilities. Additionally, operational integration and collaboration between drone operators, emergency responders, and relevant authorities need to be strengthened. As the field of drone technology continues to advance, it is crucial to address ethical considerations and ensure responsible and accountable use of drones. This involves respecting privacy rights, adhering to data protection regulations, and considering the potential impact on affected communities. Overall, drones have proven to be valuable assets in disaster management, empowering responders, and enabling more efficient and effective operations. With ongoing research, technological advancements, and collaborative efforts, the future holds tremendous potential for drones to further enhance disaster management strategies, ultimately saving lives, reducing damage, and increasing resilience in the face of disaster.

Acknowledgments

The work presented here is performed as part of the research project entitled 'Internet of Things (IoT) and Machine Learning (ML) based Smart UAV-Assisted Emergency System for Disaster Management', that is supported and funded by the Council of Science and Technology (CST), Government of UP, India (Project and Sanction No. 1199 and CST/D/1219).

Chapter 6

Trust Based Access Control in IoT Network

Rudra Krishnasrija, Sindhu Sankati, Lakshmi Sravani Popuri, Nikhitha Chava and *Amit Kr. Mandal**

The widespread adoption of IoT networks in various domains has brought about a transformative impact on our interactions with technology and the surrounding environment. Yet, as the complexity and scale of IoT deployments continue to increase, the need for ensuring network security has become a top priority. Consequently, trust-based access control has emerged as a promising security approach for IoT networks. By evaluating the trustworthiness of devices based on factors such as device behavior, historical data, reputation, and contextual information, trust-based access control offers numerous advantages. The primary objective of this project is to develop a trust-based access control mechanism specifically designed for IoT networks. By integrating traffic analysis, fuzzy clustering, and a fuzzy trust model, the project aims to assess the trust values of devices within the network. These trust values will subsequently be employed to establish access control policies, thereby enhancing the security and efficiency of IoT systems. The utilization of this trust-based approach holds great potential in enhancing reliability and secured interactions among IoT network devices while mitigating the risks associated with unauthorized access.

Department of Computer Science and Engineering, SRM University, Amaravati, India.

* Corresponding author: amitmandal@ieee.org

6.1 Introduction

The Internet of Things (IoT) networks refer to interconnected systems comprising physical devices and appliances. Some devices are furnished with sensors and connected with network, allowing them to collect, share and exchange data for various purposes with increasing proliferation of interconnected devices [14]. Traditional methods for controlling access, like access control model based on role (RBAC), discretionary access control model (DAC), and mandatory access control model (MAC), typically relies on static rules and predefined permissions, which are not well-suited for the dynamic nature of IoT systems. IoT networks are usually complex and heterogeneous. They consist of diverse devices, sensors, and communication protocols from various manufacturers. Traditional access control mechanisms often lack the capability to handle such diversity and may not support the wide range of devices and protocols found in dynamic IoT environment. To overcome these challenges, incorporating trust would be a critical factor in determining access rights and regulating interactions in IoT networks [183].

The main aim of this chapter is to create and establish a robust and efficient trust model that leverages fuzzy logic to enhance the access control capabilities of IoT networks. The suggested approach will be able to capture and reason with the inherent ambiguities and ambiguity involved in trust evaluation by utilizing fuzzy sets and fuzzy rules, enabling more nuanced and context-aware access decisions. By incorporating a fuzzy and dynamic trust model into IoT systems, our aim is to enhance the overall security, reliability, and integrity of the network. Trust becomes a fundamental factor in determining node trustworthiness, enabling effective decision-making and facilitating the identification of reliable entities within the IoT ecosystem. This work provides a thorough examination of access control in the realm of the Internet of Things (IoT).

Hence this chapter introduced an approach to enhance an improved trust access control in IoT network, combining traffic categorization, database storage, and a fuzzy trust model to assess the trustworthiness of IoT network nodes. By analyzing and categorizing network traffic, we identify patterns and behaviors, enabling us to evaluate trust levels more effectively. The resulting dataset serves as input to our fuzzy trust model, which calculates trust values for each node based on their observed traffic patterns. To assess the efficiency of our method, we performed experiments using the dataset IOT-23 [83], employing various classification algorithms. Our results indicate that our method, utilizing the Random Forest algorithm, outperforms other classifiers. This promising outcome emphasizes the potential of proposed approach to significantly improve security and enhance access control in IoT networks.

6.2 Related Works

Trust in access control plays an important role, contributing significantly to the security and reliability of IoT networks. As IoT systems continue to proliferate, establishing trust between entities is becoming increasingly essential. Consequently, the immense promise of IoT may be at risk if the concerns related to security and privacy are not adequately addressed. In this direction, numerous efforts are made to promote widespread acceptance among IoT stakeholders, thereby facilitating the wider acceptance of this evolving framework. However, when applying established authorization models like Access Control based on role (RBAC) or Access Control based on attribute (ABAC) [26], significant challenges arise due to the intricate handling of authorization policies often necessitated. This has prompted the development of solutions where smart objects operate independently of security measures and access control models. Few studies in literature recognized this issue and suggested the implementation of authorization tokens as an access control is a solution for IoT scenarios [96, 149].

Based on the ideas discussed in above studies, Hernández et al. [104] introduced a method called Distributed Capability-Based Access Control (DCapBAC) for managing access in the Internet of Things (IoT). This method works well even in situations where devices and networks possess constrained resources. Unlike traditional methods that require a central authority to decide who gets access to what, DCapBAC allows individual smart devices to control access on their own. Consequently, when interacting with IoT devices, there is no need for a central entity to oversee the communication process. But, DCapBAC lacks trust mechanisms capable of considering the reliability of the IoT devices involved in order to make decisions based on access control. In this direction, the author Chen et al. [50] have put forth a trust management model that relies on Quality of Service (QoS) metrics related to ratio of package delivery and end-to-end packet forwarding. This model utilizes fuzzy sets to handle trust and reputation but derives its findings from simulations conducted using NS-3. In a study proposed by Ben Saied et al. [215] a trust management system designed for IoT scenarios that are context-aware and multi-service in nature. This system aims to align with the specific requirements of IoT and employs various metrics to evaluate the reliability of devices with diverse hardware capabilities, accommodating the heterogeneous characteristics of IoT.

Additionally, Nitti et al. [178] introduced a method that incorporates social properties to gauge the level of trustworthiness among IoT devices. This methodology is constructed based on the Social IoT (SioT) concept, initially introduced by Atzori et al. [36]. In this framework, nodes autonomously establish social connections with one another. Atzori et al. presented two trust management models: the subjective model, where each node evaluates the trustworthiness of its peers based on its personal experiences and the opinions of their acquaintances, and the objective model. This involves the sharing of information among all devices so that they all rely on the same data. Moreover, the researchers conducted a series of simulations to assess the model's performance in the presence of potentially malicious nodes. Building on this approach, the authors Bao and Chen and Bao et al. [38, 39] put forward a dynamic trust model. This model considers various metrics that are related to social cooperation, community interests, and recommendations. Similar to earlier proposals, these studies offer both theoretical insights and simulation-based assessments. In the realm of integrating trust models into IoT access control, Chen et al. [51] introduced the notion of a Role-Based Access Control model incorporating risk awareness. This model relies solely on Quality of Service (QoS) metrics for the computation of trust values, which further assigned to specific roles. Martine et al. [154] considered fuzzy logic and proposed FRBAC, an extension and variation of RBAC based on fuzzy sets. The user-role, role permission, and consequently the assignment of the user permissions are all considered fuzzy relations by FRBAC. With FRBAC, the access decision can be defuzzified and made binary in order to handle operations that cannot be understood from a fractional perspective [154].

Furthermore, in Mahalle et al. [150], a solution called FTBAC is presented, which employs trust as a basis for access control. This system utilizes linguistic values derived from factors like experience, knowledge, and recommendations as inputs to generate values of fuzzy trust, which are subsequently associated to determine access level privileges. While simulations address scalability and energy consumption, a real-world implementation of the solution is not included. It's worth noting that these trust models share a common limitation in that they primarily rely on a limited set of parameters, such as QoS and reputation, to assess the trustworthiness and reliability of IoT devices. In recent proposals "Trust-Level-Public Key Authentication" a fresh technique for authentication has been suggested. It facilitates the connection of nodes for seamless data transfer and is backed by a cloud-based remote key management system. This system is employed to aggregate encrypted data, with remote control over the authorization levels in the nodes [235]. Garcia et al. [83] proposed a trust-based access control system for a decentralized IoT network is presented that is based on blockchain and an auxiliary TRS that incorporates trust and also reputation scores as supplementary attributes for accomplishing a trustworthy and reliable access control mechanism was designed. Another trust management system using smart contracts for decentralization is proposed by Putra et al. [195] for establishing IOT authorisation systems in the blockchain. Trust and reputation values are additional values in their access control systems that is based on attributes. A system where every access operation is documented in smart contract for enforcing access control in private blockchain networks was proposed by [195]. Moreover, this scheme lacks the ability to accommodate updates in the data structure. If modifications are made to the contract, the logic may be adjusted but the integrity of the data structure remains unaffected. Also, their framework supports IoT devices, only that possess adequate computational capabilities for executing asymmetric cryptography. Further in a multi-cloud domain, Paul et al. [191] proposed an access control framework based on trust using metrics like dishonesty level, rate of interactions, etc. Further to enhance trust in cloud service Kesarawani et al. [129] proposed cloud user-centric and cloud service provider-centric models that use fuzzy concepts to authorize and establish trust. In contrast to prior models, the approach proposed in this study comprehensively incorporates a wide range of parameters, introducing trust specifically focused on security and access control. Additionally, recognizing the inherent uncertainty in information within pervasive environments, our model employs fuzzy logic to deduce trust values. This integration aims to enhance IoT device awareness of other devices' trust scores, enabling them to adjust their access control decisions accordingly.

6.3 Trust Based Access Control in IoT Network

In IoT networks, it is crucial to establish trust for the reliable and secure operation of interconnected devices. To address this chapter introduces a comprehensive trust model that enhances the trustworthiness of individual nodes in the IoT network. Our trust model takes into consideration the unique characteristics and challenges posed by IoT systems, which are highly dynamic in nature. Therefore, to accurately reflect this dynamic environment, we incorporate the notion that trust is inherently fuzzy in nature. The general model proposed in this chapter is described in Section 6.3.1. The classifier used for traffic analysis is mentioned in Section 6.3.2. Trust calculation is described in detail in Section 6.3.4.

6.3.1 Trust Based Access Control using Fuzzy Model

For implementing our access control model, this chapter presents a framework which is summarized in Figure 6.1. Firstly, we capture traffic from different IOT devices and send it to a classifier, Which classifies the packets into broader categories, namely 'unauthorized requests', 'bad requests', and 'bogus requests'. These labels are used as arbitrary categories to group together packets that exhibit similar types of attacks or suspicious behavior.

The aim of this categorization is to streamline the analysis procedure and facilitate the recognition of patterns or shared attributes among packets within the same category. By consolidating similar attack types, it becomes more convenient to analyze their features, recognize possible threats, and develop efficient countermeasures. Following the categorization phase, the data derived from packet analysis is stored in a database. This database acts as a repository for the data collected, facilitating convenient access and retrieval of related information. The stored dataset includes traffic counts from specific nodes within the

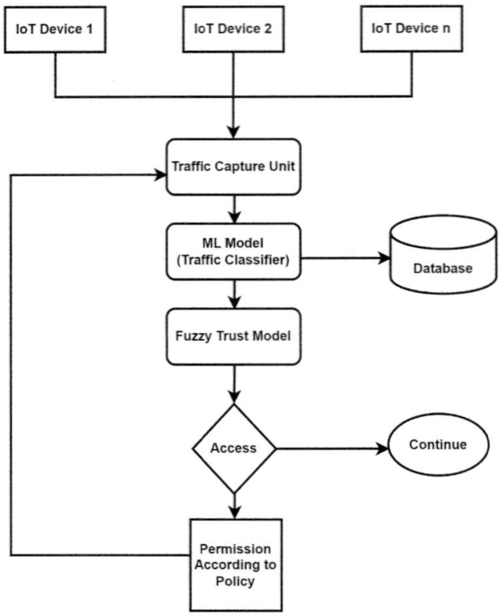

Figure 6.1: Overview of the model.

IoT network. Through the continuous tracking and recording of various traffic counts, this dataset offers valuable insights into the activities and behavior of each network node. This data serves as the foundation for further analysis in decision-making. The dataset is employed as input for the fuzzy trust model, using the stored traffic counts to compute a trust rating for each node. The fuzzy trust model evaluates a node's reliability by considering the frequency and characteristics of various types of traffic originating from it.

The computed trust value signifies the degree of trust attributed to a specific node within the IoT network. It serves as a quantitative measure of the node's reliability and integrity based on its observed traffic patterns. The trust value can be used as a basis for making access control decisions and determining appropriate actions to be taken. Furthermore, the trust value can inform the actions to be taken in response to a node's behavior. For example, if a node exhibits suspicious or malicious traffic patterns, appropriate actions such as blocking or quarantining the node may be triggered based on its trust value.

6.3.2 Traffic Analysis

According to the result of classifiers for the IOT-23 [83] dataset according to the analysis from [235], the best approach so far is the random forest algorithm as we can see from the Table 6.1 with a weighted precision of about 1 and this was

Table 6.1: Confusion matrix of classifier models IOT-23 dataset.

Metrics		Classifiers				
		RF	**NB**	**ANN**	**SVM**	**ADA**
Precision	Weighted	1.00	0.76	0.71	0.60	0.86
	Macro	0.88	0.27	0.33	0.23	0.55
Recall	Weighted	1.00	0.23	0.66	0.67	0.87
	Macro	0.85	0.38	0.14	0.14	0.35
F1 Score	Weighted	1.00	0.25	0.52	0.59	0.83
	Macro	0.84	0.10	0.10	0.13	0.37
	Accuracy	1.00	0.23	0.66	0.67	0.87

followed by an AdaBoost with a weighted precision of 0.86. A point to note is that the values are rounded to two decimal points, so the real value for random forest algorithm is not exactly 1, but it is around 0.995 to 1.000. The same applies for all the classifiers as well as for recall and $F1$ Score. Random forest uses decision trees as their meta classifier, not only this but AdaBoost also uses the same. So, we can make an assumption that the decision tree is solely needed for predicting correctly in the majority of the cases. A little better Artificial Neural Network has a tendency for those with larger numbers of occurrences compared to others. Here, we can make an assumption that it is based on the configuration of neurons with tendency in the weight functions that are not large enough which does not allow flexibility for accurate predictions. And the worst results are shown by the Support Vector Machine algorithm which has only 60% accuracy but the recall and $F1$ score are higher when compared to NB and ANN. This indicates that there are very low true positives rather than a large number of false negatives. This also shows that it is predicting the classes with more number of occurrences better. This is a confusion matrix in which only 3 classes have correct predictions but then being the biggest malware classes. So, we can see that benign labels do not have a clear understanding of various attack types, often classifying them based on similarity to the nearest-matching attack. Furthermore, Naive Bayes yields unsatisfactory results, deviating from the assumption that data points are independent, indicating a flaw in this naive assumption. In the case of the most prevalent attack categories, all approaches demonstrate accuracy rates of at least 69%, highlighting the challenge in accurately detecting attack classes with smaller categories of attacks.

Therefore instead of having all categories, in this chapter similar categories are merged together and then Random Forest algorithm is applied as observed in paper [235], we can say it performs the best in classification for the chosen dataset.

6.3.3 Trust Model

This section focuses on calculating trust values for individual nodes based on traffic categories. This calculation considers various factors, including the number of authentication failure requests (AFR), bogus requests (BOR), bad requests (BAR), and the total requests (TR) made by each node over a defined time period (T). Additionally, the implementation incorporates fuzzy inference systems, specifically utilizing the Mamdani fuzzy inference method with Gaussian membership function. To create a fuzzy model for trust, we start by clustering the data using a reference model. To conduct the preliminary clustering, we have integrated a trust calculation method, which is elaborated in Section 6.3.4. Following that, we delve into the implementation of the fuzzy interface in Section 6.3.5.

6.3.4 Trust Calculation

This section introduces an approach for determining trust values for individual nodes based on traffic category. It also outlines a method for computing trust scores for each node. The following are essential notations.

- **AFR:** Requests transmitted by a node identified as authentication failures within a specified time interval T.
- **BOR:** Requests issued by a node classified as bogus requests within a defined time frame T.
- **BAR:** Requests transmitted by a node classified as bad requests within a specific time frame T.
- **TR:** The cumulative count of requests initiated by the user within the time frame T.
- T_neg**:** The negative trust value.
- **NT:** The node trust value.

The provided authentication failure request rate is

$$AFRR = \frac{AFR}{TR} \tag{6.1}$$

Bogus request rate is calculated as

$$BORR = \frac{BOR}{TR} \tag{6.2}$$

Bad request rate is calculated by

$$BARR = \frac{BAR}{TR} \tag{6.3}$$

Trust value evaluation formula

$$T_{neg} = w1 \times AFRR + w2 \times BORR + w3 \times BARR \tag{6.4}$$

Node Trust value is given by

$$NT = 1 - T_{neg} \tag{6.5}$$

where weight factors are represented as $w1$, $w2$, $w3$ such that $w1 + w2 + w3 = 1$. Here $w1 = 0.5$, $w2 = 0.2$, $w3 = 0.3$

6.3.5 Fuzzy Inference System

The general process of fuzzy inference involves fuzzification, where crisp inputs are mapped to fuzzy sets using appropriate membership functions. These fuzzy sets are then evaluated based on predefined rules to generate a fuzzy output. Finally, defuzzification converts the fuzzy output into a crisp value, allowing for practical application and interpretation. In the proposed model, we utilize the Mamdani fuzzy inference method with a Gaussian membership function. This function operates within a range of 0 to 1 to convert the crisp input clusters into fuzzy sets. This conversion is necessary because fuzzy inference systems rely on rule-based evaluation. Consequently, we need to define rules for evaluating the outputs, which in this case, correspond to trust values as depicted in Table 6.1. The combination of these rules allows us to obtain a fuzzy output, which is a fuzzy set representing the aggregated information from the input variables. However, to make practical use of this output, we must convert it into a crisp value through the process of defuzzification. In our model, We employ a triangular membership function for determining fuzzy weights within the range of 0 and 1.

6.4 Results and Disscussion

6.4.1 Dataset

The dataset for this study is obtained from [83], the $IOT-23$ dataset, which was published by Avast AIC laboratory. The dataset comprises 23 captures, of which 20 are malicious captures and 3 benign captures. The dataset provides two kinds of download, the smaller one containing conn.log.labeled files, while the larger one also contains the p-cap files. Since for this work only the log files have been used we chose to use the former. The dataset comprises a total of $325,307,990$ captures, with $294,449,255$ of them categorized as malicious. The dataset [83] provides various attributes related to network captures such as “ts” representing the capture time, “uid” for the capture ID, and “id orig.h” indicating the IP location of the attack. Other important information includes network protocol “proto”, application-level protocol “service”, and duration of data exchange.

Additionally, it includes details on data volume, connection states, and capture labels.

For traffic analysis as discussed in Section 6.4.2, Random Forests are used in this chapter. Firstly combine all the 23 captures into one single csv file. In the original data there are two types of labels for data, one which indicates if it is benign, other more detailed label which gives which kind of attack was the packet generated. Again relabel them combining these both labels into the broad categories of 'unauthorized requests', 'bad requests' and bogus 'requests' as given in the Table 6.2.

Table 6.2: Labels of data for our classifier model to accordance to the dataset labels.

Label	Detailed Label	Classified
Benign	-	Successful
Malicious	C&C	Bad_Request
Malicious	C&C File Download	Bad_Request
Malicious	C&C-HeartBeat	Bad_Request
Malicious	C&C-HeartBeat-FileDownload	Bad_Request
Malicious	C&C-Mirai	Bogus_Request
Malicious	C&C-Torii	Bogus_Request
Malicious	DDoS	Authentication_failure
Malicious	FileDownload	Bogus_Request
Malicious	Okiru	Bogus_Request
Malicious	Okiru-Attack	Bogus_Request
Malicious	PartOfAHorizontalPortScan	Authentication_failure
Malicious	Attack	Bogus_Request

6.4.2 Traffic Analysis Observations

In this chapter, we employed the use of sklearn (version: 1.10.0) models to train our dataset, utilizing the RF algorithm. As part of the preprocessing step, We conducted a dataset partition, dedicating 70% for training and allocating the remaining 30% of dataset for testing and evaluation. To ensure the quality and relevance of the data used for classification, we excluded certain features from the analysis. Specifically, we dropped the features: "*ts*," "*uid*," "*id.orig_h*," "*local_orig*," "*local_resp*," "*missed_bytes*," *and* "*tunnel_parents*" from consideration. These features were disregarded for different reasons: the first five were considered labels rather than informative attributes for the classification task, and the last three were omitted due to a significant amount of missing data. During the training and prediction phase of our model, we observed that the process took approximately 100 seconds to complete. The model's performance was evaluated using a prediction score of 0.73, which indicates the accuracy of the model's predictions.

Table 6.3: Report on classification.

	Precision	Recall	F1-Score	Support
Authentication_failure	0.73	1.00	0.84	289445
Bad_Request	0.56	0.15	0.24	4592
Bogus_Request	0.99	0.02	0.03	79848
Successful	0.96	0.55	0.70	59518
accuracy	-	-	0.75	433403
macro avg	0.81	0.43	0.45	433403
weighted avg	0.81	0.75	0.67	433403

For further details and comprehensive information refer to Table 6.3, where additional findings and results are summarized.

6.4.3 Trust Model Observations

In our evaluation, the capabilities of the MATLAB® software were leveraged, specifically utilizing the Fuzzy Logic Toolbox, to conduct a comprehensive analysis. The objective of this chapter was to evaluate trust within a network environment. To accomplish this, we performed a traffic analysis and subsequently grouped the hosts and traffic labels together, resulting in a dataset tailored for trust evaluation. From this dataset, we selected 1000 nodes to form the basis of our analysis. To effectively classify and categorize the nodes based on their similarities, we employed the fuzzy c-means clustering algorithm. This algorithm partitions the dataset into distinct clusters, with the parameter value $c = 4$ specifying the desired number of clusters. Each cluster represents a different category of traffic within the network. The membership function after clustering is described in Figure 6.2 providing a visual description of the clusters.

Figure 6.3, Figure 6.4, and Figure 6.5 present an overview of bad requests, bogus requests, and unauthorized requests relative to the total number of requests. The proposed fuzzy model that we have developed can be further expanded to encompass additional categories of traffic beyond bad requests, bogus requests, and unauthorized requests. This versatility allows for the evaluation of various types of traffic in diverse IoT networks, making our model highly adaptable and applicable to a wide range of scenarios. The experimental analysis when the data is reflected in a control surface shows that in our dataset, with an increase in total requests that there were less bad requests and authorised failure thus having high trust values as we can observe in Figure 6.3 and Figure 6.4. While the bogus requests have been a little more complicated, since the high total requests had a lot of bogus requests thus having a less trust value, therefore there is bulge in the control surface as in Figure 6.5.

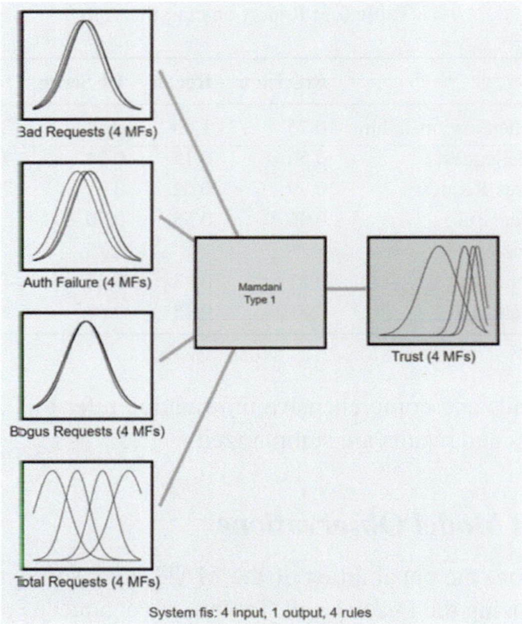

Figure 6.2: Heat map.

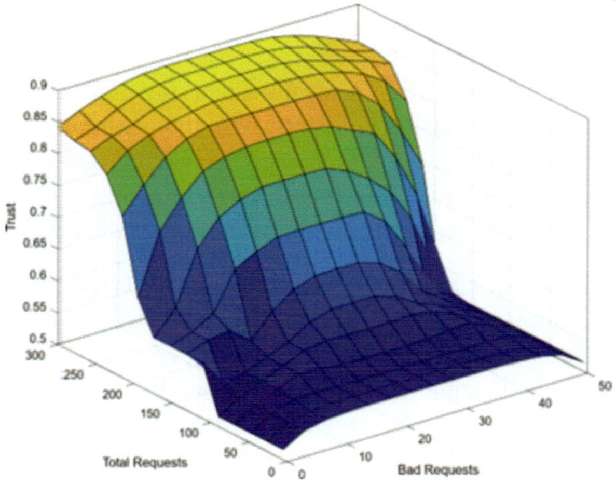

Figure 6.3: Control surface view of total requests and bad request.

The proposed model results in a root mean square error (RMSE) of 0.0251. By leveraging this model, we obtain higher trust values for good users compared to the mathematical model. Similarly, for malicious users, the trust values are lower than those determined by the mathematical model.

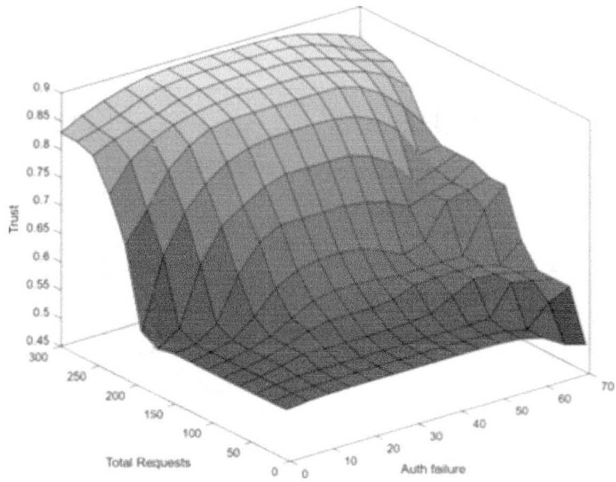

Figure 6.4: Control surface view of total requests and auth failure request.

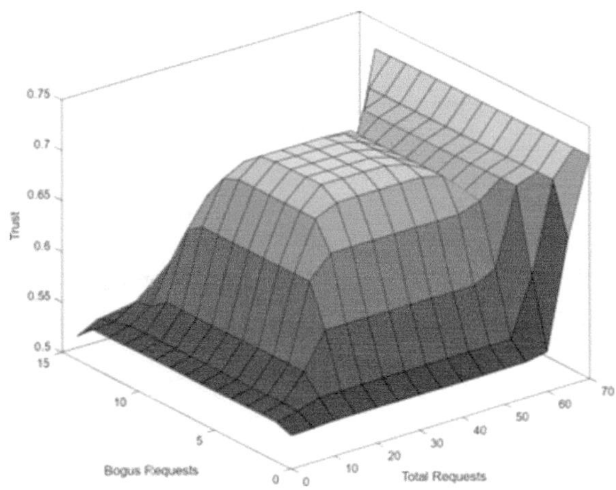

Figure 6.5: Control surface view of total requests and bogus requests.

6.5 Concluding Remarks

In closing, the goal of this project was to introduce a distinctive trust-based access control mechanism in order to address the security challenges in IoT networks. By utilizing methods like traffic analysis, fuzzy clustering, and a fuzzy trust model, we were able to construct a framework that could assess the trustworthiness of devices and set up access control rules. We learned important things about the various traffic patterns that particular hosts inside the IoT network generate through an extensive analysis of network traffic. We successfully classified related traffic patterns into informative clusters using the fuzzy c-means (FCM) clustering technique. Our capacity to distinguish between distinct device behaviors was much improved through the clustering method, allowing for more precise trust judgements. Using a fuzzy trust model with clustered traffic patterns, each device in the network was successfully assigned a trust value. With the use of these trust values, we were able to discern between trustworthy and potentially dangerous devices by using accurate assessments of device credibility and reliability. We also proposed using the generated trust values as the foundation for developing access control rules. These regulations are essential for controlling how devices communicate with one another and limiting access to vital resources based on how reliable each device is judged to be. The importance of this research rests in its contribution to resolving issues related to access control and trust in IoT networks. But it's important to acknowledge that there's still opportunity for development and investigation. Future research may concentrate on improving the fuzzy trust model by taking into account more criteria for judging trust. Additionally, analyzing the framework's performance in actual IoT settings would offer crucial insights for its useful application. In conclusion, the project's trust-based access control method makes a significant addition to the IoT security sector. We want to improve the security and dependability of IoT networks by offering an innovative approach that incorporates trust ratings into access control mechanisms.

6.6 Future Work

The utilization of fuzzy trust models in IoT access control offer a number of promising directions for future research. Firstly, we still need to explore alternative membership functions, rule aggregation, and inference mechanisms in order to improve and enhance the accuracy and reliability of trust evaluations. Additionally, investigating dynamic trust adaptation is crucial, as it involves developing methods to adjust the fuzzy trust model based on real-time contextual information, such as device behavior and user preferences, to accommodate evolving trust dynamics. Future research can concentrate on developing methods that allow trust evaluation while protecting the privacy of sensitive data. Privacy-preserving trust computation is another key direction for future work. Furthermore, addressing scalability and efficiency challenges in trust computation, particularly in large-scale IoT deployments, is a crucial area for future investigation, involving distributed trust evaluation mechanisms and optimization techniques. Real-world evaluation and validation through extensive experiments, field trials, and case studies are also necessary to assess the effectiveness, robustness, and applicability of the proposed fuzzy trust models in diverse IoT scenarios and applications.

Chapter 7

Safeguarding User Privacy at the Edge of Fog Computing Networks in Decentralized Distributed Computing

Chandramohan Dhasarathan,[1] *Ramachandra Reddy B.,*[2,*] and *Diwakar Tripathi*[3]

Establishing secure identities and authentication measures is crucial to thwart unauthorized access to fog nodes and edge devices. Fog computing environments offer enhanced user privacy protection while leveraging the advantages of decentralized and distributed computing at the network's edge. Directly conducting data analytics on edge devices without transmitting raw data to the cloud minimizes the risk of sensational sensitive evidence during the data processing phase. Incorporating rigorous access mechanism contrivances ensures that solitary ratified individuals can enter sensitive data, employing role-based control supervision skills. Additionally, homomorphic encryption is advocated, as it en-

[1] Thapar Institute of Engineering and Technology, Patiala, Punjab, India.
[2] National Institute of Technology Jamshedpur, Jharkhand, India.
[3] Indian Institute of Information Technology, Sonepat, Hariyana, India.
* Corresponding author: brcreddy.cse@nitjsr.ac.in

ables computation on encrypted statistics without the prerequisite for decryption, facilitating protected data dispensation while preserving privacy. The escalating pervasiveness of IoT diplomacies and growing demand for low-latency applications, fog-based research focus has arisen as a auspicious paradigm that encompasses the cloud computing capabilities of network operations. Nevertheless, the decentralized and distributed nature of fog computing raises substantial concerns regarding the privacy and security of user data. This research article explores various strategies and techniques to safeguard user privacy within the context of decentralized and distributed computing in fog computing networks. The examination encompasses encryption, anonymization, access control, and other privacy-preserving mechanisms designed to address the distinctive challenges posed by the fog computing environment.

7.1 Introduction

The adoption of fog computing in contemporary computing architectures has ushered in improved performance and decreased application latency. Nevertheless, the decentralized and distributed characteristics of fog computing bring forth new challenges, particularly in the realm of safeguarding user privacy. This study aims to examine and develop practical strategies to address the challenges associated with safeguarding user data confidentiality in fog computing environments. In the Internet of Things (IoT) era and the increasing demand for swift-response communications, fog computing has emerged as an innovative framework, acting as a bridge between the centralized cloud and the decentralized edge. Extending computational capabilities to the network's edge through fog computing presents unique prospects for improved performance and real-time processing [13].

Nevertheless, this paradigm's decentralized and distributed nature gives rise to fresh challenges, particularly concerning confidential data. The intrinsic architecture of fog computing, reliant on interconnected nodes and edge devices, necessitates thoroughly exploring strategies to preserve user privacy. Establishing secure identity and authentication mechanisms is a fundamental cornerstone in thwarting illicit entrance to fog nodes and edge devices. Implementing strong access controls and employing role-based methods becomes essential to guarantee that only authorized entities can engage with sensitive data [239, 138].

This research advocates for a further shift in data analytic practices to fortify user privacy. Performing analytics directly on edge devices without transmitting raw data to the cloud becomes a pivotal strategy. This not only reduces latency but also mitigates the inherent risk of exposing sensitive information during data processing. In tandem with these strategies, incorporating homomorphic encryption introduces a transformative layer of security. In the processing phase, this encryption technique enables computations on encrypted data without decryption, ensuring data privacy preservation. The confluence of these measures establishes a framework wherein fog computing environments can harness the benefits of decentralized and distributed computing at the edge without compromising user privacy. This research article delves into an in-depth exploration of these privacy-preserving strategies. From encryption and anonymization to access control mechanisms and homomorphic encryption, our investigation aims to discuss the distinctive challenges impersonated by the fog computing environment. By dissecting these strategies, we endeavour to contribute to the evolving landscape of privacy protection in the decentralized and distributed computing paradigm at the edge of fog computing networks [201, 23].

7.2 Privacy Dilemmas in Fog Computing Landscape

Privacy challenges in fog computing are paramount due to this emerging computing paradigm's matchless appearance and architecture. Several factors contribute to the significance of addressing privacy concerns from the fog computing perspective. Fog computing's decentralized structure involves distributing computing resources through a network of knots, including edge devices. This dispersion raises the jeopardy of unauthorized access and data exposure, making robust privacy measures essential. Fog computing emphasizes processing data at the edge near the data source. While this enhances efficiency and reduces latency, it also poses challenges in securing sensitive information during on-site data processing. Privacy safeguards become critical to prevent data leaks [251, 203].

Fog computing operates in heterogeneous networks, incorporating diverse devices and communication protocols. Managing privacy across such environments requires adaptable strategies to secure data transmission and storage. With the explosion of IoT maneuvers connected to fog computing networks, the time and variety of data being processed have expanded significantly. Safeguarding user privacy is essential as these devices often handle sensitive personal information. Fog computing facilitates real-time data processing, a necessity for autonomous vehicles and industrial automation applications. Ensuring privacy in such environments is crucial to prevent unauthorized access and potential misuse of real-time data. Transmitting data between devices over potentially insecure networks introduces risks of interception and eavesdropping. Encryption and secure communication protocols are essential to mitigate these risks and protect user privacy. Privacy challenges in fog computing are compounded by the need to comply with data protection guidelines such as GDPR, HIPAA, and others. Non-compliance poses legal risks and erodes user trust in supervising their search information [227, 79].

Addressing privacy challenges is crucial for building and maintaining user trust. Users are increasingly concerned about the security of their data, and fog computing providers must establish robust privacy measures to encourage the widespread adoption of these technologies. Beyond legal requirements, ethical considerations surround the responsible handling of user data. Privacy challenges in fog computing must be tackled to ensure data is processed and utilized ethically, respecting user rights and expectations. As technology evolves, so do cyber-security threats. Privacy encounters in fog computing need to be continuously addressed to stay ahead of incipient extortions and vulnerabilities, safeguarding user data against potential breaches. The importance of managing privacy challenges in fog computing lies in fostering a secure and trustworthy user environment, complying with regulations, and ensuring the data's responsible and ethical use in an increasingly interconnected and decentralized computing landscape. The unique challenges related to user privacy in fog computing include data transmission over heterogeneous networks, data processing at the edge, and the potential exposure of sensitive information [6, 138].

7.3 Literature Study

In the study [197], Rahman et al. have explored current research that centres around emerging technologies such as Blockchain (BC), Software Defined Networking (SDN), and Smart Industrial Internet of Things (IIoT). They underscore the contributions of these technologies to data integrity, confidentiality, and integrity, with a specific emphasis on their applications in industrial settings. The paper acknowledges the long-standing use of cloud computing, but notes associated security risks. To address these challenges, the authors propose a cloud computing platform named "DistB-SDCloud" that integrates Blockchain and Software Defined Networking for enhanced security in intelligent IoT applications. The design utilizes a distributed Blockchain to ensure security, confidentiality, privacy, and integrity, all while preserving flexibility and scalability. Software Defined Networking aims to improve the durability, stability, and load balancing of cloud infrastructure. The paper conducts experimental tests on the proposed implementation, assessing parameters such as throughput, packet analysis, response time, bandwidth, latency, and the system's resilience against attacks.

Xue et al. [251] discuss the evolving landscape of Internet of Things (IoT) devices, noting the shift from traditional cloud-enabled IoT frameworks to a bidirectional network framework with the advent of edge computing. In this bidirectional framework, IoT devices are capable of performing computational tasks, transforming them from mere data harvesters to active computing entities. However, the shift to edge computing introduces new security risks, particularly when data and models are transferred back to IoT devices. To address these risks, the researchers propose Acies, a privacy-focused classification system built on differential privacy principles. Acies ensures the security of classification models delegated to edge devices and supports widely used classifiers such as Nearest

Neighbourhood, Support Vector Machine, Sparse Representation Classifier, and various feature selection methods.

Ajao et al. [13] discuss the challenges posed by the vulnerability of Industrial Internet of Things (IIoT) networks in smart cities to malicious attacks and cyber threats. The study proposes a security framework using Petri Net to address these vulnerabilities. The framework introduces a standard trust model with a distributed authorization algorithm (STAPPA) to manage information outflows, focusing on privacy breach mitigation. They primarily optimize search, detect anomalies, and determine the shortest route during agent learning. The simulation results indicate a high detection and accuracy rate of 98.75%, 99%, 99.50%, 99.75%, and 100%, with an average sensitivity of 1.000. Al-Huthaifi et al. [203] explores the development of smart cities (SC) globally over the past decade, emphasizing the use of big data and the Internet of Things (IoT) to enhance efficiency, livability, and sustainability by monitoring and integrating various infrastructure systems. However, the increased reliance on big data makes smart cities more susceptible to cyber-attacks, posing challenges to privacy and security. With growing regulations on user data protection, the article discusses the application of federated learning (FL) as a solution to enhance privacy. FL, a decentralized machine learning approach, distributes learning across diverse datasets, offering pluses for delicate and dissimilar data dominions in smart cities. The study comprehensively reviews FL's application in dissimilar SC classifications, such as conveyance, healthcare, and communication. It explores the benefits and drawbacks and identifies technological problems and risks associated with FL implementation. The article highlights the need for further comprehensive testing of existing FL systems against various attacks to improve data protection and performance in intelligent cities, emphasizing the importance of ongoing research in this area. Keshari et al. [130] focus on vehicular fog computing (VFC) in their article, aiming to enhance intelligent traffic systems (ITS) by processing real-time traffic information. They utilize it to issue accident alerts and navigate paths. VFC utilizes parked and moving vehicles as fog nodes for computation, effectively decreasing bandwidth consumption, response time, and congestion in the core cloud owing to shorter distances. The system conducts resource allocation to select the suitable vehicle for computation, considering the available computational resources. It conducts data salvage maneuvers to transference circulation statistics from vehicles to vehicular fog nodes. However, VFC encounters challenges such as high vehicle mobility, dynamic topology, inept supply portion, and late data delivery. Security and privacy concerns arise from sharing traffic data within vehicular fog. To tackle these challenges, the article suggests conducting an in-depth examination of each segment of VFC operation.

Quan et al. [196] suggest a framework based on Blockchain to overcome challenges in outmoded cloud-centric curative statistics partaking. The focus is on addressing real-time recital. The framework leveraged healthcare consortium edge Blockchains in edge computing environments to provide valid admittance mechanism to curative data. It tackles explicitly endorsement challenges through a scheme established. Tabassum et al. [239] explore the field of the Internet of Vehicles (IoV), specifically focusing on vehicular ad hoc networks (VANETs). It discusses the challenges associated with traditional cloud-centric and decentralized edge computing approaches, emphasizing latency and infrastructure costs. The emergence of integrating automatic vehicles to the virtual environment is introduced as a solution to these challenges, impacting areas such as transportation. VANETs, known for their quick handoffs, network reliability, and security features, play a crucial role in IoV-CC. Despite the benefits of cloud computing (CC), concerns about quality of service (QoS) parameters led to the VANET that supports fog computing. The research aims to explore data distribution and security acceptability in IoV-CC, particularly in the context of centralized and decentralized computing. It underscores gaps in data dissemination and security challenges in innovating the Quality of Service in VANET for IoV-CC. The article concludes by discussing ongoing efforts, consequences, and potential future directions for overcoming these challenges. Singh et al. [227] discuss the potential of blockchain technology in addressing security concerns related to access control in Internet of Things (IoT) devices. Blockchain's distributed and decentralized design is a promising solution for ensuring secure and scalable transactions among IoT devices. Previous studies have explored the use of Blockchain in IoT to enhance security. The abstract highlights that traditional access control mechanisms may need to be revised to effectively meet the requirements of large-scale IoT systems in terms of manageability, scalability, and accuracy in protecting hardware and data. The study examines the potential of a blockchain-based decentralized architecture to improve access management in IoT, aiming to provide a more effective solution for ensuring security in the context of IoT devices.

Frimpong et al. [79] discuss the challenges related to privacy in recommendation systems, particularly those who wish to be unknown on online platforms like open medium and e-commerce. Concerns arise from commercializing users' data, leading to conviction disputes concerning user data providers who may sell data to third parties without user consent. The framework prototype was implemented in its experimental results demonstrate the effectiveness and privacy of users within the proposed scheme. RecGuard aims to give online users control and autonomy over their data, addressing the privacy challenges associated with recommendation systems in online social networks. Basset et al. [6] address the cyber security challenges smart cities face with the increasing dominance of Internet of Things (IoT) schemes. It focuses explicitly on privacy-related threats such as inference and data poisoning attacks within these complex networks. Federated Learning (FL) is considered a promising solution for distributed learning while preserving privacy in IoT applications. However, current inquiry primarily ponders on FL with independently identically distributed (i.i.d) data, neglecting the non-i.i.d setting. The authors introduce a framework to tackle privacy challenges by deep-learning approaches. It aims to protect data against privacy-related Generative Adversarial Network (GAN) attacks and achieve high classification rates in the context of non-i.i.d data. Private identifiers generated for each class are used to protect class probabilities. The proposed framework holds promise for enhancing privacy in federated deep learning for IoT applications in smart cities.

The Khezri et al. [138] article addresses the challenge of managing copious massive facts generated by Internet of Things (IoT) devices. While cloud computing platforms traditionally handle such data, specific applications face unbearable delays due to data transfer edge components to the virtual storage, especially when numerous devices are connected, causing the public network bottleneck. This situation requires a supplementary vigorous frame and complex processes for supremacy, data stowage, supply controlling, and overhaul administration. To enhance the resourceful use of grid and cloud possessions, the article explores the capabilities of fog computing, particularly its "intelligent gateway" feature. The focus is on planning and managing resources, which significantly impacts particularly the latency environment. Lee et al. [144] address challenges in industrial applications where IoT devices are pivotal for continuous and extensive data collection but encounter limitations in storage capacity. However, Interplanetary File System (IPFS) has drawbacks, including time inefficiencies in data recording and difficulties conveying source device identity and meta-data through Content Identifiers (CIDs). This complexity impedes efficient condition-based data retrieval and prevents effective storage management by IoT devices. IoT devices can use the archived metadata to determine data status, facilitating effective storage management by removing unnecessary data. The proposed architecture is versatile, long suffering data of unpredictable sizes, layouts, and regularities. Shruti et al. [201] introduce the concept of Industry version 5.0, an evolving sector which uses human collaboration with advanced technologies and aims to transform work processes, mainly contributing to the improvement of insolent conurbations. The focus is designing and operating intelligent activities of IIoT devices, involving intricate decision-making at different levels. Real-time records congregation from IoT maneuvers is crucial. Fog Computing is highlighted for its role in positioning expedients close to their vicinity, sinking treating time compared to cloud computing. Fog computing is emphasized for enabling intelligent systems' secure, flexible, and time-sensitive operation. Despite challenges, including data-intensive analysis due to sensor presence, fog computing is presented as an effective solution. Security is addressed by introducing a categorized fog computing-based style for innovative energy-supplying organisms within the framework of Industry version 5.0. This architecture is designed to support numerous infrastructural components efficiently.

Alsaeed et al. [23] discuss security challenges in the IoMT, emphasizing the inadequacies in the computation, storage, and self-protection capabilities of IoMT devices. While recent blockchain-based authentication schemes offer robust security, they suffer from high reckoning, storage requirements, and long endorsement stretches for medical technology systems. The paper recommends a unique group substantiation structure for IoMT systems to address these issues. Simulations on the Ethereum platform using Solidity language demonstrate that the proposed framework achieves an average latency of 0.5 seconds for authenticating IoMT devices and a throughput of 400 transactions per second. Comparative analysis against other blockchain-based authentication techniques reveals superior performance in terms of latency and throughput. The framework's average latency remains consistent within a narrow range, indicating scalability support for IoMT systems even with increasing submitted transactions. Nandhakumar et al. [170] highlight the growing significance of edge computing in addressing the needs of next-generation IoT applications, focusing on minimizing latency and optimizing bandwidth utilization. Despite its benefits, edge computing faces challenges in resource management, prompting researchers to explore AI models as solutions. Current simulation tools, however, need to adequately support the integration of AI mockups for reserve supervision, presenting a hurdle for researchers developing novel policies. Javanmardi et al. [118] focus on fog computing for supporting time-critical IoT applications at the network's edge, emphasizing the need for security and efficient scheduling in IoT fog networks. It highlights the vulnerability of SDN switches and controllers, often serving as fog gateways, to various attacks, posing challenges to security and control plane saturation. Comprehensive simulations demonstrate that S-FoS outperforms existing algorithms in IoT-based scenarios. Experiment results show notable improvements in response time, under varying attack rates, the number of IoT devices, and fog devices.

Ding et al. [63] address the challenge of automatically detecting epileptic seizures in an Internet of Medical Things (IoMT) environment using electroencephalogram (EEG) signals. Traditional approaches involve collecting EEG signals from different patients and centralizing them on a server, risking patient privacy and incurring energy and communication costs. Dhasarathan et al. [59] focus on COVID-19 health analysis and estimate using tolerant data from sanatoriums and health officialdoms. It highlights the potential risk to users' credentials and personal information, emphasizing the need for a secure approach to minimize such risks globally. The article proposes a homomorphic-based identification system to detect possible breaches, aiming to enhance privacy preservation in various fields. The complexity of maintaining secret patient information, especially with health data collected from diverse scenarios, is acknowledged. The article advocates for a systematic approach combining homomorphic techniques with deep learning processes to reduce unauthorized access and illegal functionality by unknown organizations. Additionally, the paper introduces a homomorphic privacy metric amalgamation for improvising all approaches in addressing complex healthcare system issues. The effectiveness of the proposed homomorphic-based privacy observation metric is highlighted through result analysis, demonstrating its gradual improvement in enhancing the deep learning process for COVID-19 healthcare management. Dhasarathan et al. [60] use deep learning in e-healthcare applications which has shown significant benefits for diagnosis and treatment, but concerns about patient data privacy have arisen. A new approach addresses these concerns by introducing auto-agent privacy metrics for social healthcare systems. This method combines deep learning with a multi-agent system to create a secure framework. The multi-agent system oversees data access by assigning roles and permissions to each agent, and real-time privacy metrics (confidentiality, integrity, availability) assess the data privacy level, flagging potential violations. Integration of a deep learning model, trained on extensive medical records, improves diagnostic accuracy based on patient symptoms and history. Advantages of this approach include enhanced e-healthcare security, authorized data access, real-time privacy violation identification, reduced data breach risks, and improved diagnostic precision, leading to better patient outcomes.

Razaque et al. [206] introduce an algorithm for the Mobile Fog-Based Cloud (MFBC) in the context of the energy based Internet of Things (IoT). EESH leverages a voltage scaling factor to minimize energy consumption, proving more effective than existing algorithms as the quantity of IoT errands increases. Moreover, the algorithm is enhanced with security features, incorporating a different algorithm that supports blockchain to safeguard user identities. The processors on the server side of MFBC are optimized for IoT tasks. The proposed algorithms, implemented in Java, are evaluated against other algorithms, demonstrating superior performance in all metrics. Dhasarathan et al. [58] address privacy concerns in healthcare monitoring within the IoMT. It introduces a comprehensive solution by integrating Blockchain, homomorphic encryption, and supervised learning-based intelligent contracts. This approach, called Blockchain for IoMT (BIoMT), leverages distributed ledgers to manage IoMT data, ensuring privacy and security in clinical information. BIoMT employs amalgam encryption techniques, enhancing patient and health record privacy protection. The system also incorporates a secure and bearable hoard series managing framework using blockchain-based smart contracts. The proposed BIoMT method demonstrates superior effectiveness in preserving privacy and preventing data breaches through testing and comparison with existing prevention mechanisms. Murugaiyan et al. [165] address critical issues in cloud storage services regarding handling users' private info and ensuring its confidentiality. The focus is on preserving user data privacy in cloud computing, as the information stored in the cloud is susceptible to various security threats and privacy breaches. The framework aims to prevent unauthorized access, intrusions, and breaches of confidential digital storage by implementing data de-identification, log monitoring, and maintaining a secure cloud environment. The goal is to enhance privacy protection for providers and users of cloud storage services.

7.4 Proposed Encryption Technique for Distributed Fog Computing

Homomorphic encryption serves as a cryptographic method allowing computations on encrypted data without decryption, ensuring secure data processing within fog computing environments. In the context of fog-enabled critical infrastructure, where maintaining confidentiality, integrity, and availability is crucial, effective key management and secure communication are essential components of any system. A centralized essential management authority generates and securely stores cryptographic keys. This authority generates vital pairs (public and private keys) for different encryption schemes, such as homomorphic and symmetric/asymmetric encryption. Securely distribute cryptographic keys to fog nodes and edge devices. Use secure channels and protocols for key exchange to prevent eavesdropping or interception during the distribution process. Implement mechanisms for key revocation in case a fog node or edge device is compromised or decommissioned. This ensures that the compromised entity cannot decrypt or access sensitive information. Periodically rotate cryptographic keys to enhance security. Key rotation reduces the impact of a potential key compromise and aligns with best practices for maintaining a robust security posture. Make certain that cryptographic keys are securely stored on fog nodes and edge devices. Employ hardware security modules (HSMs) and trusted execution environments (TEEs) to safeguard key material against unauthorized access. Evaluate the significance of end-to-end encryption and encryption at rest in maintaining the confidentiality of user data. Explore the application of homomorphic encryption for secure computation without compromising privacy in fog computing environments.

Algorithm 8 Homomorphic encryption technique for distributed Fog computing environment (Function Definitions)

From phe import paillier
Generation of public and private keys for the homomorphic encryption scheme public-key, private-key
= pallier.generate-paillier-keypair();
Function-1 to encrypt data using the public key:
def encrypt-data(data, public-key):
encrypted-data = [public-key.encrypt(x) for x in data];
return encrypted-data;
Function-2 to perform homomorphic addition on encrypted data: def homomorphic-addition(encrypted-data-1, encrypted-data-2):
result = [x + y for x, y in zip(encrypted-data-1, encrypted-data-2)];
return result
Function-3 to perform homomorphic multiplication on encrypted data
def homomorphic-multiplication(encrypted-data, scalar):
result = [x * scalar for x in encrypted-data];
return result;
Function-4 to decrypt homomorphically processed data using the private key def decrypt-data(encrypted-result, private-key): decrypted-result = [private-key.decrypt(x) for x in encrypted-result];
return decrypted-result;

7.4.1 Privacy-Preserving Data Analytics

The following section will examine the direct execution of data analytics on edge devices to minimize reliance on cloud-based data transmission. It aims to evaluate the concept of differential privacy as a means to optimize data utility while ensuring individual privacy during analytics. This section stress the imperative of obtaining explicit user consent before initiating data collection and processing. It will investigate methods for enhancing transparency, empowering users to control their privacy settings. The aim is also there to assess the compliance of fog computing networks with relevant data protection laws and standards. We also intend to analyze the impact of regulations on privacy preservation within fog computing and illustrate with real-world case studies, demonstrating successful implementations of privacy-preserving strategies in fog computing environments.

Algorithm 9 Transaction of fog node receives encrypted data from IoT devices

Step 1: Encrypt data received from IoT devices
data-from-iot = [5, 8, 12, 3];
encrypted-data = encrypt-data(data-from-iot, public-key);
Step 2: Perform homomorphic operations (addition and multiplication)
For example, adding a constant to the encrypted data
constant-to-add = 10;
encrypted-result-addition = homomorphic-addition(encrypted-data, encrypt-data([constant-to-add] * len(data-from-iot), public-key));
Computing the encrypted data by a scalar multiplication process
scalar-to-multiply = 2;
encrypted-result-multiplication = homomorphic-multiplication(encrypted-data, scalar-to-multiply);
Step 3: Decrypt the homomorphically processed result
decrypted-result-addition
= decrypt-data(encrypted-result-addition, private-key);
decrypted-result-multiplication = decrypt-data(encrypted-result-multiplication, private-key);
Display the results:
print("Original Data:", data-from-iot);
print("Encrypted Data:", encrypted-data)
print("Decrypted Result (Addition):", decr;ypted-result-addition);
print("Decrypted Result (Multiplication):", decrypted-result-multiplication);

7.4.2 Transport Layer Security (TLS)

The Algorithms 8 and 9 implements TLS for secure communication between fog nodes, edge devices, and other components in the infrastructure. TLS ensures data confidentiality and integrity during transmission. It enables mutual authentication to verify the identities of both communicating parties. This prevents man-in-the-middle attacks and ensures that fog nodes and edge devices communicate with trusted entities. It uses digital certificates for authentication and establishing the trustworthiness of fog nodes and edge devices. To maintain a secure certificate management system to issue, renew, and revoke certificates as needed. Adopting secure communication protocols for specific use cases within the critical infrastructure is a must. The implementation of end-to-end encryption to protect data integrity and confidentiality from the source to the destination. This is particularly crucial for sensitive information flowing through the fog computing infrastructure.

7.4.3 Anonymization Strategies Adapted for Fog Computing

To establish a continuous security auditing and monitoring system to swiftly detect anomalies or potential security breaches, covering unauthorized access, unusual communication patterns, and changes in key usage we need to employ anonymization strategies. This can be implemented by employing intrusion detection and prevention systems that actively monitor network traffic and behavior to identify and address potential security threats. The discussion extends to the application of anonymization techniques for safeguarding user identities and sensitive data during both transmission and storage. Figure 7.1 underscores the critical balance between preserving data utility and the anonymization process. By seamlessly integrating these anonymization strategies into fog computing systems, organizations can effectively navigate the delicate balance of utilizing valuable data for analytics while ensuring the privacy of individuals contributing to that data. It is crucial to customize these strategies based on specific use cases, regulatory requirements, and the inherent nature of the data processed at the edge.

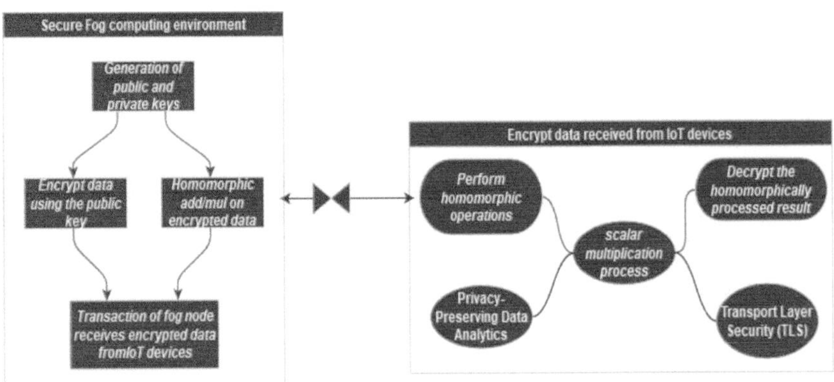

Figure 7.1: Architecture flow of fog computing for privacy-preserving.

A quantitative basis for evaluating different aspects of the authentication and authorization model. The actual metrics and calculations will depend on the specific goals and requirements of the fog computing environment and critical infrastructure in question.

7.4.4 Pseudonymization

Pseudonymization is the process to replace personally identifiable information (PII) with pseudonyms or tokens. This ensures that the data can still be used for analytics and processing without revealing the actual identities of users. Fog nodes can handle the pseudonymization process locally.

7.4.5 K-Anonymity

Implementing k-anonymity by grouping individuals into clusters of at least k individuals with similar characteristics. This reduces the likelihood of re-identification. Fog nodes can perform clustering locally before transmitting aggregated, anonymized data to the cloud.

7.4.6 L-Diversity

The Algorithms 8 and 9 extend the k-anonymity by ensuring each cluster has diverse sensitive attribute values. This prevents homogeneity within clusters, making it more challenging to infer specific information about individuals. Fog nodes can enforce l-diversity before forwarding data.

7.4.7 Differential Privacy

Utilizing differential privacy techniques involves introducing noise to data before analysis, thereby reducing the risk of identifying specific individuals while retaining the ability to derive meaningful insights. Localized fog nodes can actively contribute to the noise injection process in this approach.

7.4.8 Dynamic Masking

Implementation of dynamic masking to continuously change or update anonymization techniques. This adaptability helps thwart potential attacks that may exploit static anonymization methods. Depending on the evolving threat landscape, fog nodes can periodically adjust the anonymization mechanisms.

7.4.9 Location-Based Privacy Techniques

Edge devices leverages location-based privacy techniques to obfuscate or generalize location data. This is particularly relevant in fog computing scenarios where edge devices may transmit location information. Fog nodes can locally transform precise locations into generalized or obfuscated representations.

7.4.10 Tokenization

Edge computing are using tokenization to replace sensitive data elements with randomly generated tokens. This method protects sensitive information like credit card numbers or medical records. Fog nodes can tokenize data before transmission to enhance privacy.

7.4.11 Secure Multi-Party Computation (SMPC)

Implementation of SMPC to allow multiple parties to jointly compute a function over their inputs while keeping those inputs private. Fog nodes can contribute to the computation without revealing the raw data to each other or the cloud.

7.4.12 Hashing Techniques

By employing secure hashing techniques to convert sensitive data into irreversible, fixed-size hash values. This practice proves valuable for safeguarding both integrity and anonymity. Fog nodes have the capability to perform local hashing before transmitting data, enhancing the security of the process.

7.4.13 Privacy-Preserving Aggregation

Privacy-preserving aggregation techniques to perform computations on encrypted and anonymized data without revealing individual values. This allows nodes to transmit aggregated results instead of raw data to the cloud.

7.4.14 Consent Management

Implementing consent management systems that allow users to specify the level of anonymization they are comfortable with. This user-centric approach empowers individuals to have control over the degree of privacy they desire in fog computing scenarios.

7.4.15 Access Control and Authorization for Fog Computing

Access control and authorization are crucial to securing fog computing critical infrastructure, ensuring that only authorized entities can access sensitive resources and perform specific actions. We are examining the implementation of robust access control mechanisms, such as role-based access control (RBAC), to restrict unauthorized access to user data and define the role of secure identity management in ensuring the integrity of fog computing networks. By integrating robust essential management practices with secure communication protocols, fog-enabled critical infrastructure can establish a fast data processing, transmission, and storage foundation. This holistic approach helps mitigate the risks associated with potential security threats in critical systems.

7.4.16 User Authentication

The process involves implementing robust user authentication mechanisms to verify the identity of users accessing the fog computing infrastructure. It utilizes multi-factor authentication (MFA) for an additional layer of security, requiring multiple forms of identification. It can integrate with existing identity providers to use standard authentication protocols such as OAuth and OpenID Connect for centralized authentication. OAuth and OpenID Connect-based authentication and authorization models provide a robust and standardized approach for securing access to resources within a fog computing critical infrastructure. It establishes a foundation for secure and scalable identity management and access control in a decentralized and distributed computing environment. Access tokens secure communication between fog nodes, edge devices, and protected resources, ensuring that only authorized entities can access critical infrastructure resources.

7.4.17 Role-Based Access Control (RBAC)

Establish distinct roles for various users and entities within the fog computing infrastructure based on their respective responsibilities. Assign specific permissions to each role, outlining the actions or resources that users with that role can access. Conduct regular reviews and updates of role assignments to ensure alignment with organizational changes and security policies.

7.4.18 Attribute-Based Access Control (ABAC)

Employ ABAC to make access control decisions based on attributes of users, devices, and other contextual information. Framing well-structured policies that consider various attributes such as user roles, location, time, and device characteristics to determine access rights dynamically.

7.4.19 Fine-Grained Access Control

Introduce precise access controls to limit user access at the level of individual resources or actions. This approach ensures that users are granted the minimum required permissions for their tasks, effectively minimizing the risk of unauthorized access. OAuth and OpenID Connect are scalable and widely supported, facilitating interoperability between different components of the fog computing infrastructure. OpenID Connect includes user consent mechanisms, ensuring users explicitly grant permission to share their identity information.

7.4.20 Secure Device Authentication

Establish a secure authentication mechanisms for fog nodes and edge devices to ensure that only authorized and trusted devices can participate in the infrastructure. Use device certificates, fast boot processes, and secure communication protocols to authenticate and authorize devices. Access tokens obtained through OAuth provide a safe and scalable mechanism for controlling resource access.

OAuth and OpenID Connect enable centralized authentication, allowing users and devices to authenticate through a single, trusted identity provider.

7.4.21 Access Control Lists (ACLs)

Implement ACLs to define and enforce rules for access to specific resources or services. Regularly audit and update ACLs to reflect infrastructure changes and maintain an accurate representation of authorized entities. The fog node can use the identity token to retrieve additional user information from the OpenID Connect. The Reserve Server certifies the access token with the OAuth 2.0 Authorization Server to ensure its authenticity and permissions. Upon successful authentication and user consent, the OAuth 2.0 Authorization Server gives an access token to the fog node.

7.4.22 Centralized Policy Management

Centralize the management of access control policies to ensure consistency and ease of administration. Use a policy controlling classification that allows administrators to define, modify, and enforce policies across the entire fog computing infrastructure. The OpenID Connect IdP authenticates the user and obtains their consent to release identity information to the requesting fog node. The fog node sends an authorization request to the OAuth 2.0 Endorsement Headwaiter to get access to a safe reserve on behalf of the user. Users access a fog node or edge device and initiate the authentication process. The fog node redirects the user to the centralized OpenID Connect Identity Provider (IdP) for authentication.

7.4.23 Logging and Auditing

Implement comprehensive logging mechanisms to capture access attempts, changes to access control policies, and other relevant security events. Regularly audit logs to detect and respond to potential security incidents or policy violations. Represents the server hosting protected resources, applications, and data within the fog computing environment. Represent the entities within the fog computing infrastructure with which users and devices interact. This serves as the identity provider responsible for authenticating users and providing identity information. Extends OAuth 2.0 to include an authentication layer and additional user information.

7.4.24 Dynamic Access Policies

Consider dynamic access policies that can adapt to changing conditions in real time. This includes adjusting access rights based on contextual changes, user behavior, or security events. It acts as the centralized authorization server responsible for issuing access tokens. It manages client registration and defines access policies for protected resources.

7.4.25 Regular Security Assessments

Regularly perform security assessments, encompassing penetration testing and vulnerability assessments, to pinpoint and rectify potential vulnerabilities in the access control and authorization mechanisms. Users, administrators, and devices must access and interact with resources and services within this critical infrastructure. OAuth and OpenID Connect will be used to ensure secure and centralized authentication.

7.4.26 Compliance with Regulations

Ensure access control mechanisms align with relevant regulations and standards, such as GDPR, HIPAA, or industry-specific compliance requirements. OAuth and OpenID Connect (OIDC) are widely adopted protocols for authentication and authorization in modern systems, providing a secure and standardized way to manage access to resources.

7.4.27 Incident Response Planning

Designing an incident response plan that includes procedures for addressing unauthorized access or breaches in the access control system. Regularly test and update the incident response plan to enhance its effectiveness. To safeguard sensitive resources and data from unauthorized access and misuse, organizations can establish a secure and resilient environment by implementing strong access control and authorization mechanisms in fog computing critical infrastructure. It's essential to adapt and customize this framework based on the specific needs and characteristics of the critical infrastructure.

7.5 Result Analysis

In this study various measures for comparative analysis are as defined in Equations 7.1–7.9. The result analysis process should involve a combination of quantitative metrics, qualitative feedback, and a comprehensive evaluation of the proposed authentication and authorization model. Regular reviews and updates to the model may be necessary based on the findings of the result analysis process to ensure ongoing effectiveness and alignment with evolving security requirements.

$$\begin{aligned} &Hash-based\,Message\,Authentication\,Code(HMAC), Access\,Token \\ &\qquad = HMAC(Client\,Secret, Data\,to\,be\,Signed) \end{aligned} \tag{7.1}$$

$$Probability(User\,Consent\,Granted)(UCP) = \frac{Number\,of\,Consents\,Granted}{Total\,Authentication\,Attempts} \tag{7.2}$$

$$Access\,Control\,Accuracy(ACA) = \frac{Number\,of\,Correct\,Accesses}{Total\,Number\,of\,Access\,Attempts} * 100\% \tag{7.3}$$

$$Throughput(AT) = \frac{Number\,of\,Authentications}{Time\,Taken} \tag{7.4}$$

$$Rate\,of\,Adaptation(RA) = \frac{Number\,of\,Dynamic\,Policy\,Changes}{Total\,Time\,Duration} \tag{7.5}$$

$$Vulnerability\,Index(VI) = \frac{Number\,of\,Vulnerabilities}{Total\,System\,Components} * 100\% \tag{7.6}$$

$$User\,Satisfaction\,Score(USS) = \frac{Positive\,User\,Feedback}{Total\,User\,Feedback} * 100\% \tag{7.7}$$

$$Incident\,Response\,Time(IRT) = \frac{Time\,Taken\,to\,Respond\,to\,Incidents}{Number\,of\,Incidents} \tag{7.8}$$

$$Scalability\,Index(SI) = \frac{Number\,of\,Users\,or\,Devices\,Added}{System\,Performance\,Impact} \tag{7.9}$$

They can recommend future improvements, scalability adjustments, and areas where the model can be refined for sustained effectiveness. As well as assess the system's ability to withstand security challenges and recover from incidents, and contribute. Table 7.1 shows the OAuth-OpenID Connect model consistently outperforms the traditional model across various metrics. Higher authentication success rates, improved user consent probabilities, and enhanced access control accuracy contributes to a more secure and user-friendly environment. Improved scalability, faster incident response, and dynamic policy adaptation highlight the advantages of the OAuth and OpenID Connect-based model in fog computing.

To assess whether users perceive the system as user-friendly and privacy-aware and whether they feel in control of their identity information. We can examine how well the model aligns with security best practices and if it effectively addresses security concerns associated with fog computing environments. It should

Table 7.1: Result comparison table for the proposed authentication and authorization model for fog computing.

Metric	OAuth-OpenID Connect Model	Traditional Model	Remarks
ASR	98%	95%	Higher success rate with OAuth
UCP	90%	85%	Improved user consent with OpenID Connect
ACA	97%	92%	Enhanced accuracy with OAuth tokens
AT	500 auth/sec	300 auth/sec	Higher throughput with OAuth
RPA	10 changes/min	5 changes/min	Faster adaptation with dynamic policies
VI	2%	5%	Lower vulnerability with OAuth
USS	92%	88%	Improved user satisfaction with OpenID
IRT	15 minutes	30 minutes	Faster incident response with OAuth
SI	95%	85%	Improved scalability with OAuth

Authentication Success Rate (ASR)

be possible to assess how well the system adapts to evolving conditions in real time, such as changes in user behavior or alterations in the fog computing environment. This assists in determining if the system provides sufficient audibility and accountability to detect and respond to unauthorized access attempts or security breaches.

Table 7.2 compares metrics between the OAuth-OpenID Connect model and the SAML-Based model in the context of authentication and token-related performance. OAuth-OpenID Connect exhibits higher throughput, indicating a faster authentication process. It demonstrates faster token generation, completing the process in 5 ms. Moreover, it results in lower latency for resource access, with tokens being processed more efficiently.

Table 7.2: Performance metrics OAuth-OpenID Connect model for fog computing.

Metric	OAuth-OpenID Connect Model	SAML-Based Model	Remarks
AT	500 auth/sec	400 auth/sec	Higher throughput with OAuth
TG	5 ms	10 ms	Faster
RAL	15 ms	20 ms	Lower latency with OAuth tokens

Token Generation with OAuth (TG) # Resource Access Latency (RAL)

Table 7.3 presents the OAuth-OpenID connect model and it exhibits better security metrics, with a lower vulnerability index, faster incident response times, and stronger token encryption than an API key-based model.

Table 7.4 presents the OAuth-OpenID connect model and it excels in privacy and user-centric metrics, showcasing higher user consent probability, satisfaction scores, and support for privacy-preserving aggregation compared to an LDAP-based model.

Table 7.3: Security Metrics for secure Fog computing.

Metric	OAuth-OpenID Connect Model	API Key Model	Remarks
VI	2%	8%	Lower vulnerability with OAuth
IRT	15 minutes	45 minutes	Faster incident response with OAuth
TES	AES-256	AES-128	Stronger encryption with OAuth tokens

Token Encryption Strength (TES)

Table 7.4: Privacy and user-centric Metrics.

Metric	OAuth-OpenID Connect Model	LDAP-Based Model	Remarks
UCP	90%	75%	Improved user consent with OpenID Connect
USS	92%	80%	Higher user satisfaction with OAuth
PPA	Enabled	Not Supported	Enhanced privacy with OAuth

Privacy-Preserving Aggregation (PPA)

In Figure 7.2 the bar chart visually compares the authentication success rates between the OAuth-OpenID Connect and traditional models. Figure 7.3 illustrates the difference in user consent probabilities between the OAuth-OpenID Connect and an alternative model. Figure 7.4 compares vulnerability indices between the OAuth-OpenID Connect model and an API key-based model. Similarly, Figure 7.5 compares incident response times between the OAuth-OpenID Connect and legacy authentication models. The plots show the user data authentication process and visually represent critical metrics and comparisons related to the proposed authentication and authorization model.

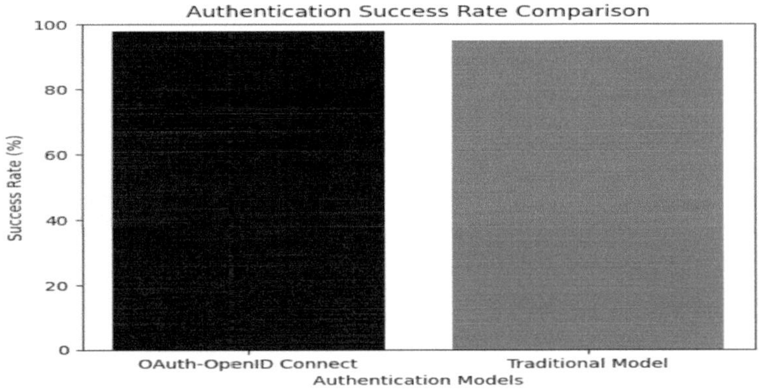

Figure 7.2: Authentication success rate comparison graph.

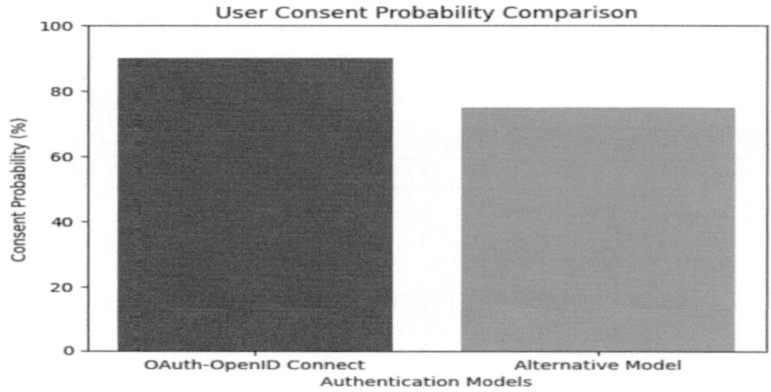

Figure 7.3: User consent probability graph.

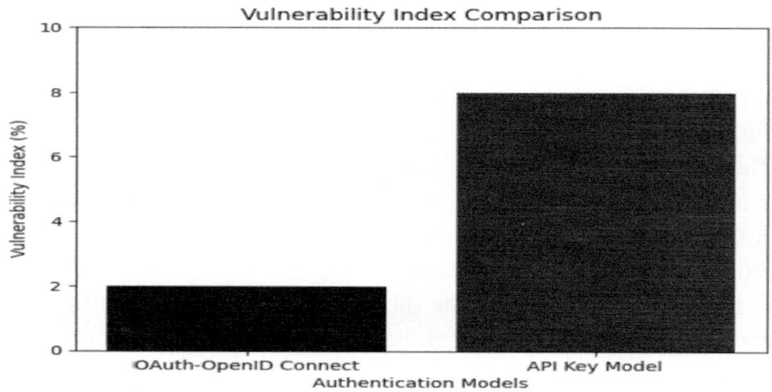

Figure 7.4: Vulnerability index graph.

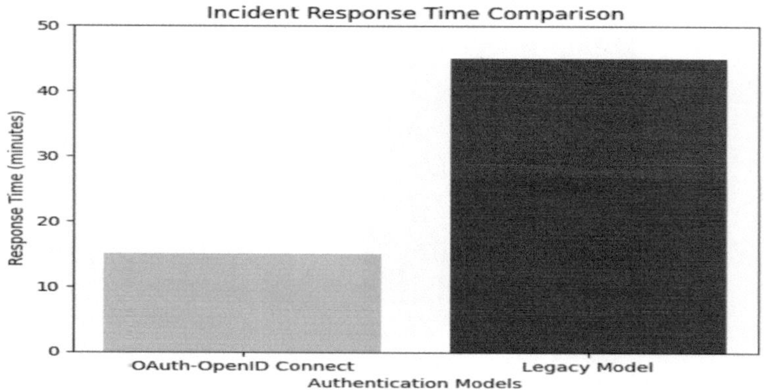

Figure 7.5: Incident response time graph.

The observations highlight the research's focus on enhancing fog computing environments' overall security and privacy. By combining standardized protocols, user-centric approaches, and fine-grained access control, the proposed model aims to establish a robust security posture. The research provides a clear and concise work-flow illustration, aiding stakeholders in understanding the sequence of events in the authentication and authorization process. This clarity supports effective implementation and ongoing maintenance of the proposed model. The article hints at the potential for dynamic access policies, acknowledging the need for adaptive systems in fog computing environments. This adaptability is crucial in environments where conditions and user interactions are dynamic and unpredictable. Token validation and centralized logging mechanisms contribute to auditability and accountability. This is essential for monitoring security events and facilitating timely responses to potential security incidents. It recognizes the security benefits of using access tokens to secure resource access. Access tokens serve as bearer tokens, providing a secure and efficient means of controlling access to protected resources within the critical infrastructure. Appropriate Leveraging OAuth-based access tokens allows for fine-grained access control. The model ensures that fog nodes and edge devices only possess the necessary permissions, enhancing security by limiting access to specific resources.

7.6 Conclusion

The research shown within this chapter demonstrates a thoughtful and comprehensive approach to addressing authentication and authorization challenges in fog computing critical infrastructure. By leveraging standardized protocols, emphasizing user consent, and ensuring scalability, the proposed model is well-positioned to enhance the security and privacy of fog computing environments. Ongoing research and practical implementations based on these principles will likely contribute to the maturation of secure fog computing practices the summarization of the essential findings and contributions of the study. As well as emphasize the importance of privacy preservation in decentralized and distributed computing at the edge of fog computing networks, and propose future research directions in the field. By addressing the challenges and implementing the proposed strategies, fog computing can realize its potential while ensuring the privacy and security of user data in an increasingly interconnected and edge-centric computing landscape.

Chapter 8

Attack Resilient Authentication Scheme for Fog-centric IoT Device Systems

Irfan Alam,[1,*] *Zubair Ashraf*[1] and *Chour Singh Rajpoot*[2]

Internet of Things (IoT) devices are growing exponentially. Exponential growth in IoT devices makes it an integral part of today's world. Fog computing is one of the effective technologies to provide various application support to match with its fast-growing speed and user satisfaction in terms of efficient utilization of resources. However, the inclusion of fog computing also brings new security problems in the form of various attacks. To use fog-centric IoT devices we propose an attack-resilient authentication scheme with the check of correctness of user inputs. Proposed schemes are designed with special attention to the low latency, robustness and low storage capacity of the IoT devices. Formal security analysis with AVISPA simulator verifies its security standard. In addition, Informal analysis shows it's resistant to many attacks which makes the proposed scheme more acceptable for Internet of Things to Internet of Every Things (IoET).

[1] SCAI, VIT Bhopal University, Kothrikalan, Sehore, M.P., India-466114.
[2] SCOPE, VIT Bhopal University, Kothrikalan, Sehore, M.P., India-466114.
* Corresponding author: irfanalam@vitbhopal.ac.in

8.1 Introduction

The Internet of Things (IoT) has a big impact on our daily lives. It turns ordinary objects into intelligent ones by allowing them to interact and synchronize with one another. This facilitates data exchange and decision-making [15]. There is a lot of data gathered from IoT devices, thus processing and storing the data requires a suitable platform. Cloud servers have been utilized in order to meet this need [174, 17]. When combined with the cloud, IoT technologies will make it easier to complete tasks because of their enhanced capacity to handle and store data [64]. However, there are still certain issues. There will be significant latency, difficulty locating the IoT device, and impossibility of uniformly dispersing the devices over a specific area. The IoT-cloud scenario is useless for deployment in a variety of application domains due to the aforementioned shortcomings. IoT is combined with fog computing to overcome the aforementioned shortcomings by expanding the network, storage, and computing capabilities to the network edge [174].

The primary goal of fog computing is to not only offer the advantages of cloud computing but also to bridge the gaps between remote data centers and Internet of Things (IoT) devices. Additionally, it brings about improvements in security, reduces computational overhead, minimizes bandwidth usage, and results in comparatively lower latency, as highlighted by [245, 119]. Consequently, fog computing emerges as a promising technology for various IoT services such as smart homes, smart cities, healthcare, vehicular networks, and more, as indicated by Wazid et al. [247]–[199]. In addition to these benefits, fog computing facilitates timely communication and supports mobility through local awareness. Since real-time data from smart devices in the IoT are accessed by numerous users, privacy and security issues inherited from cloud computing, such as spoofing, impersonation, insider attacks, password guessing, and user anonymity, must be addressed in the fog computing environment, as pointed out by Wazid, Adhikari, and Sadhukhan [247, 9, 200].

As a result, addressing these security concerns necessitates the implementation of a robust user authentication and key management scheme for IoT communication, ensuring secure operation over both secure and insecure channels within the fog computing environment. Consequently, only authorized and authenticated users would be granted access to the system and valuable information. This paper conducts an in-depth analysis of numerous recent studies on authentication and key management specifically tailored for the context of fog computing. A summary of our contributions are as follows:

- Detailed literature review of security (authentication) fog-centric IoT systems.

- A novel authentication scheme is proposed with the concentration of lightweights.
- Informal security analysis to prove the proposed scheme is attack resilient to various attacks.
- AVISPA implementation for formal security verification.

8.2 Literature Review

The effective management of dynamic ownership data in a fog environment holds significant importance. In 2018, Koo et al. introduced an authentication scheme, as outlined in their work [141], which not only addresses this aspect but also ensures the preservation of owner's privacy and the de-duplication of data. Subsequently, Wang et al. [245] presented a secure data aggregation scheme designed for a fog-enabled public cloud. This scheme collects data from fog nodes, terminals, or edge nodes and securely transmits the aggregated data to the cloud server. Following that, in 2019, Sowjanya et al. [232] introduced a lightweight IoT communication scheme based on Elliptic Curve Cryptography (ECC). This scheme employs key-policy attribute-based encryption (KP-ABE) along with key refresh and key update techniques. In the same timeframe, Li et al. [146] presented a secure three-factor user authentication scheme, leveraging Elliptic Curve Cryptography (ECC), specifically designed for wireless medical sensor networks. This scheme ensures forward secrecy, with biometric information required for authentication managed through a fuzzy commitment scheme.

Furthermore, in 2019, Harbi et al. [97] proposed a scheme tailored for Wireless Sensor Network (WSN)-based IoT applications. The system does not provide secure mutual authentication or session key agreement, despite its claims to be safe against replay, denial of service, and impersonation attacks. During this period, Wajid et al. [247] introduced a three-factor scheme tailored for the fog-centric IoT communication environment. This scheme, named SAKA-FC, efficiently manages cryptographic keys and authenticates different entities involved in secure communication within a fog computing environment. The three factors employed in this protocol are the user password, biometrics, and the mobile device. The authors utilized a fuzzy extractor technique for biometric verification using the user's mobile device. Additionally, cryptographic one-way hash functions and XOR operations are incorporated into the protocol to ensure lightweight implementation, making it suitable for devices with limited resources.

In 2020, Sadhukhan et al. [214] introduced an Elliptic Curve Cryptography (ECC)-based remote user authentication scheme employing three-factor verification. This approach aims to safeguard user privacy and data confidentiality against various attacks. Simultaneously, using an ECC-based authentication mechanism, Rostampour et al. [211] demonstrated an effective way to secure communication between cloud servers and IoT edge devices. Moving forward to 2021, Wu et al. [249] conducted a comprehensive analysis of the scheme proposed by Jia et al. [119] and identified vulnerabilities to various attacks. As a response, they proposed an authentication and key agreement scheme specifically designed for IoT healthcare systems using fog computing. However, it was noted that Wu et al.'s scheme had high communication overhead, impacting overall efficiency. In the realm of fog-based authentication for vehicular ad-hoc networks, Wu et al. [248] presented another scheme where fog nodes serve as relay nodes for mutual authentication. For lightweight authentication in the context of Internet of Drones, Jan et al. [116] proposed a scheme addressing long-standing issues such as insider attacks and stolen verifier attacks.

8.3 Preliminaries

8.3.1 Elliptic Curve Cryptography

In 1985, Koblitz [140] and Miller [158] introduced the definition of the discrete logarithm problem on a set of points of an elliptic curve. The unique characteristics of the elliptic curve group, which lacked a natural analogue for the Index Calculus method, made it the preferred choice for Koblitz and Miller. Despite attempts by other mathematicians, such as Jacobsen and others [113, 180, 226], to extend the Index Calculus for elliptic curves, these efforts proved unsuccessful. Following the proposal by Koblitz and Miller, Elliptic Curve Cryptography (ECC) gained popularity. Similar to RSA, ECC operates as an asymmetric key mechanism. ECC is particularly efficient for devices with constrained resources due to its ability to provide enhanced security with lower power consumption, reduced computational overhead, and smaller key sizes compared to other algorithms like RSA or DH, as highlighted by Paulo [40].

Elliptic Curve Cryptography (ECC) not only contributes to memory and bandwidth savings but also integrates the discrete logarithm problem [67, 179] and ElGamal's cryptographic algorithm to formulate the ECC algorithm. ECC's mathematical operations are conducted on points of an elliptic curve, and its strength is contingent on the size of the keys employed for encryption [189]. Moreover, ECC presents certain advantages over cryptographic methods operating in integer fields. In comparison to other cryptographic algorithms like RSA and Diffie-Hellman [61], which function in integer fields, ECC employs significantly smaller key sizes. This reduction in key size is a distinctive feature of Elliptic Curve Cryptography. Furthermore, the mathematical foundation of ECC relies on the complexity of the Elliptic Curve Discrete Logarithm Problem (ECDLP) [67].

It is theoretically challenging to discover an integer n such that $Q = nP$ if there are two points $P, Q \varepsilon E(p)$ (where $E(p)$ is an elliptic curve), according to the

ECDLP. The difficulties of ECDLP is taken into account in our suggested approach. In the event where $Q = s.P + t.P$ and $P = \varepsilon E(p)$ are two points such that $Q = s.P + t.P$, then finding points $s.P$ and $t.P$ across $E(p)$ is mathematically challenging [252]. The term elliptic curve factorization problem (ECFP) describes this difficulty. The intractable nature and complexity of the Elliptic Curve Discrete Logarithm Problem (ECDLP) and the Elliptic Curve Factorization Problem (ECFP) are prerequisites for the efficacy of the suggested strategy.

8.3.2 Bilinear Pairing

If G1 is an additive group, G2 is a multiplicative group and P is a generator of G1, then Mapping $\hat{e} : G_1 \times G_2 \rightarrow G_2$ is called a bilinear pairing, where G_1 and G_2 are additive group and multiplicative group respectievely and P is generator. It satisfies the following properties:

- Bilinearity: For any $P, Q, R \varepsilon G$. $\hat{e}(P+Q, R) = \hat{e}(P, R).\hat{e}(Q, R)$.Therefore, for any $a, b \varepsilon Z^* : \hat{e}(aP, bQ) = \hat{e}(P, Q)^{ab} = \hat{e}(abP, Q) = \hat{e}(P, abQ)$ holds.
- Non-degenerate: For any $P \varepsilon G_1, \hat{e}(P, P) \neq 1$, where 1 is the identity element.
- Computability: There is an efficient algorithm to compute $\hat{e}(P, Q)$ for any $P, Q \varepsilon G_1$.

G1 is defined over an elliptic curve, G2 is a multiplicative subgroup and the mapping $\hat{e}$ will be obtained from the modified Weil pairing or Tate pairing [221, 18].

8.4 Proposed Framework

In this section, we introduce our secure integrated framework designed for Fog-IoT systems.

8.4.1 System Model

In Figure 8.1, the general architecture of a Fog-IoT system is depicted, comprising three essential components: IoT devices, Trusted Authority (TA), and Fog Nodes. The Internet of Things devices layer includes a range of smart devices, such as smart meters, weather monitoring systems, refrigerators, garage doors, health monitoring devices, lights, and thermostats, among others. Network infrastructure such as switches, routers, bridges, gateways, and base stations are included in the fog nodes layer. These nodes have been endowed with additional computational power and incorporate local servers, including mobile phones, embedded servers, industrial controls, and security cameras. Fog Nodes

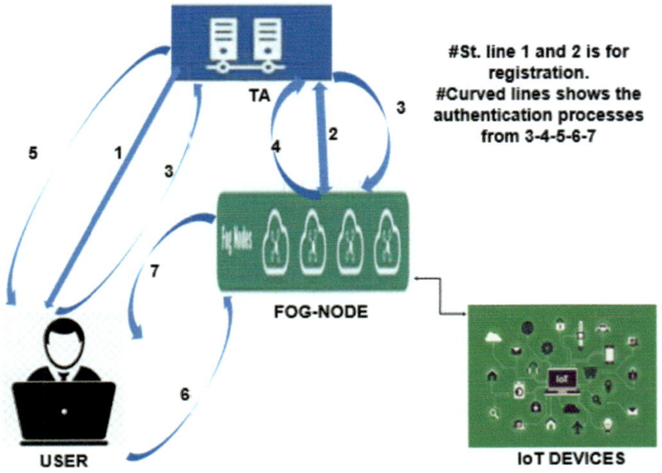

Figure 8.1: Proposed scheme.

also oversee the IoT systems. In Figure 8.1, the registration phase is denoted by st. line 1 and 2. Curved lines 3 to 7 are used for authentication purposes.

The Trusted Authority (TA) serves as a centralized computing and storage platform, offering diverse applications for acquiring, processing, presenting, and managing Fog-IoT systems. The TA's capacity is not constrained, providing a robust foundation for system functionalities. As it is shown in the figure that user and Fog node interacts with TA for thier registration in first step. In consecutive stages, TA controls their authentication process. Finally, fog node authenticates the user for access to the IoT devices.

8.4.2 Threat Model

In the considered threat model for Fog-IoT entities, three categories are identified: honest, semi-honest, and dishonest. The cloud is characterized as honest but curious, meaning it adheres to protocol specifications while gathering information about services, users, and locations. Fog nodes (FO) and IoT devices fall into the semi-honest category, implying they can be compromised and are susceptible to security and trust attacks. This model discusses many security threats. The confidentiality and integrity of unencrypted communications can be jeopardized by interception and modification. Fog nodes that have been compromised could be used as a platform for botnet attacks such as data theft, probing, distributed denial of service, and denial of service. These attacks affect availability by preventing reputable fog nodes from offering services, which makes them appear unreliable to IoT devices.

Multiple exploited IoT devices may be able to register under the same identity (Sybil attack) and change access rights thanks to malicious fog nodes. In the Fog-IoT system, trust computation is important for every entity. However, compromised nodes might distort the trust model by providing incorrect parameters, which makes the trust credibility assessment necessary. Malicious bot nodes may report fake parameters in self-promotion attacks to increase their trust, and false parameters about peers in bad-mouthing attacks. Similar techniques can be used by IoT devices to undermine fog nodes' trust. Fog nodes that provide good or bad Quality of Service (QoS) depending on their resources are the target of opportunistic and intermittent attacks [19]. In collusion assaults, adversaries build a botnet to change the target nodes' level of trust. A fog node may generate several false IDs in a Sybil attack in order to report parameters that have been made up. When a device provides more parameters than is allowed in a certain amount of time, it is known as ballot stuffing. Because of the imprecise trust computation caused by these attacks, the nodes' actual performance is contradicted, making efficient detection and mitigation strategies for trust computation necessary [122, 15].

Table 8.1: Symbols used in proposed scheme.

Symbols	Abbreviation
G_1, G_2 and , G_3	Multiplicative cyclic Group
Z_P	Large prime integer
g_1, g_2 and g_3	generators
SK	Session Key
U_i	Device ID
B_1, B_2	Public Key
TA	Trusted Authority

8.4.3 Proposed Scheme

This section provides an overview of the proposed scheme, segmented into various phases such as the pre-deployment, registration, and authentication phases. The utilization of multiplicative cyclic groups linked with Type-A elliptic curves, as defined in the Pairing-Based Cryptography (PBC) Library [4], is a key aspect of the scheme. The primary components of the proposed scheme include the IoT device, User, and Trusted Authority (TA), as detailed in the network model. During the authentication phase, a session key is established and subsequently employed for maintaining confidentiality. The outlined presumptions for the scheme are as follows:

- Fog-nodes (IoT devices), Users, and TAs have the capability to execute similar cryptographic operations.

- The TA is connected to the main power system, experiencing no computational or storage limitations. Conversely, Users and Fog-nodes operate with constrained resources.
- The Trusted Authority (TA) is considered trustworthy and immune to tampering.
- Messages are conveyed through a public channel post-registration, while a secure channel is employed during the registration process.

8.4.4 Pre-deployment Phase

To uphold the security of the system, the Trusted Authority (TA) generates the initial system settings through the following steps and continually updates them during this phase.

- Multiplicative cyclic group G_1, G_2 and , G_3 are generated based on elliptic curve.
- TA picks a random number $t \varepsilon Z_q^*$.
- TA picks a random number $a \varepsilon Z_q^*$.
- TA calculates $X_1 = g_1^a$ where g_1 is the generator of the cyclic group G_1.
- TA picks a random number $t \varepsilon Z_q^*$.
- TA picks a random number $a \varepsilon Z_q^*$.
- TA calculates $X_1 = g_1^a$.

8.4.5 Device (Fog-node) Registration

- Fog node sends device id and generator (U_i, g_2) to TA.
- TA computes private key $B_1 = g_1^{t+U_i}$.
- TA computes public key $B_2 = g_1^{a+t-U_i}$.
- TA also calculates Authentication parameter $U_1 = g_2^{\frac{1}{a+2t}}$ and send $U_2 = h(U_1 || U_i)$ to Fog node.

8.4.6 User Registration

- A particular user D_i, sends his id along with generator (d_i, g_3) to TA.
- TA computes $D_1 = g_1^{a+d_i}$, $D_2 = g_1^{a-d_i+t}$ and $D_3 = g_3^{\frac{1}{2a+t}}$.
- TA sends D_3 to particular user D_i.

8.4.7 Authentication

- Fog-node generates a nonce $U^1 \varepsilon Z_P$ send U^1 and U_2^* to TA.
- $TS_u - TS_{TA} \leq \Delta T$ and $U_2^* = U_2$.
- The Trusted Authority (TA) initiates the process by generating a random number T^1. Subsequently, it calculates T_u using the formula $T_u = h(B_1||B_2||U^1||T^1)$ and sends the result to the user.
  Simultaneously, the user transmits D^1 to the Trusted Authority (TA). The TA then calculates T_D using the formula $T_D = h(D_1||D_2||D^1||T^1)$ and forwards the result to the Fog-node.
  The user also computes $SK_D = (U^1 \oplus T^1) \oplus D^1$ and sends it to TA.
  When the user intends to access a specific IoT device, it is required to communicate with the Fog-node through the Trusted Authority (TA). The TA computes SK_{TA} using the expression $SK_{TA} = (D^1 \oplus T^1) \oplus U^1$. If SK_D is equal to SK_{TA} (indicating that the user has been authenticated by the TA), then the user is granted permission to access the Fog-node.
  The users provide (B_1, B_2) to fog-node, fog-node check or authenticate the user by:
  $e(B_1 \times B_2, U_1) = e(g_1, g_2)$
  also the fog node calculates $SK_u = \left(T^1 \bigoplus U^1\right) \bigoplus D^1$.
  fog-node Checks $SK_u = SK_D$ and $SK_u = SK_{TA}$ (fog node authenticates the trusted authority and user) After this step the fog-node sends access-key information (BI) of the Iot device to the user. Similarly, Fog-node sends T_D to the user, the user authenticates particular fog node producing the following result:
  $(D_1 \times D_2, D_3) = e(g_1, g_3)$.
  finally session key $SK_u = SK_D = SK_{TA}$ is established for further uninterrupted communication.

8.5 Security Analysis

The proposed security scheme is passed through AVISPA as per the requirement of formal security analysis (shown in the Figure 8.2).

To show that the proposed scheme is attack resilient from various attacks. The next subsection is dedicated to the informal security analysis.

8.5.1 Informal Security Analysis

- **Anonymity and Un-traceability:** During the process of exchanging keys between a fog node, a user, and a trusted authority, the protocol ensures that the identities of the involved parties remains confidential. Fresh nonces are generated independently by the fog node, the user, and the

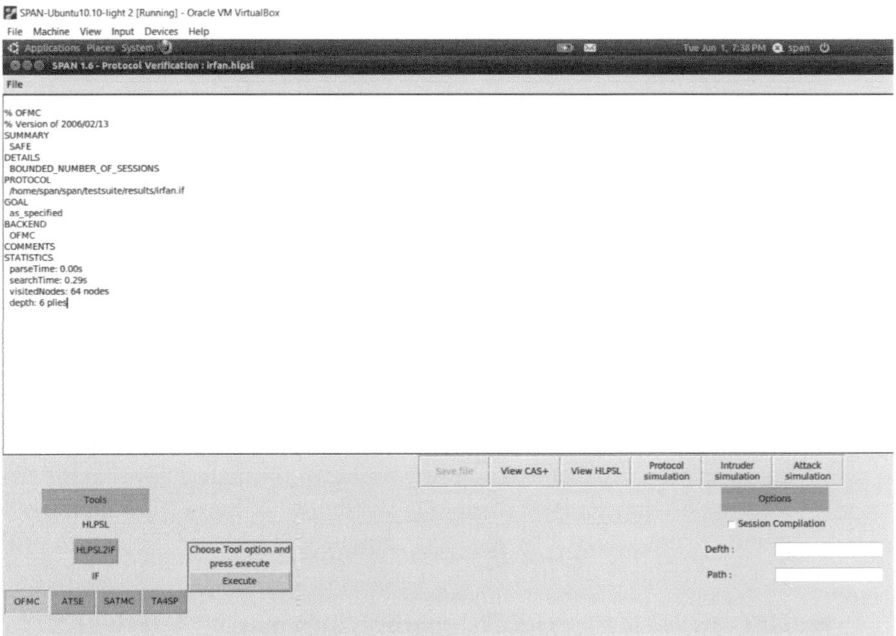

Figure 8.2: AVISPA result for proposed scheme.

trusted authority. During the registration phase of the fog node and the user, the trusted authority calculates B_1, B_2, B_3, and U_2 for the fog node using the group generator based on the elliptic curve. Similarly, D_3 is generated for the user. This ensures that a third party cannot obtain information about the identities of the fog node or the user. Therefore, the proposed protocol guarantees user anonymity and un-traceability, as the identities of the fog node and user remain confidential and cannot be revealed to external entities.

- **Offline Guessing Attacks (OGA):** Differential power analysis attacks [156] pose a potential threat for extracting information from devices like smart cards and data carriers. In the proposed protocol, the parameter U_2 is generated by hashing together two words derived using an elliptic curve generator. The hash function's one-way collision resistance property, combined with the uniqueness of Elliptic Curve Cryptography (ECC), makes it highly challenging to discern a user's identity. The consecutive and similar phases performed during registration further mitigate the risk of power analysis attacks, as these attacks are not feasible due to the protocol's design [45].
- **Perfect Forward Secrecy (PFS):** In the proposed protocol, adversaries may potentially acquire some keys and parameters. Perfect Forward Se-

crecy (PFS) ensures that all session keys between fog nodes, users, and trusted authorities remain confidential even if certain keys or parameters are compromised. In this context, consider an adversary attempting to compute $SK_u = SK_D = SK_{TA} = (D^1 \oplus T^1) \oplus U^1$. Since these parameters are consistently used in hashed situations, they remain unavailable to the attacker. The problem of finding a polynomial-time solution to obtain these parameters becomes challenging. Thus, the proposed protocol effectively provides Perfect Forward Secrecy.

- **Sybil Attack (SA):** An adversary launches a Sybil assault by generating several accounts or nodes in an effort to take over the network. The Trusted Authority's (TA) elliptic curve and its generators are used in the proposed protocol during the registration and first authentication step. The Elliptic Curve Discrete Logarithm Problem (ECDLP) in ECC has an intrinsic property that makes it impossible to compute a key secretly from known parameters. As a result, the possibility of harm from multiple accounts is not affected by the keys' uniqueness because the design of the protocol reduces the risk of Sybil assaults.

- **Known Session-specific Temporary Information Attack (KSTIA):** The opponent has access to the ephemeral session secrets in the context of the KSTIA attack, which enables them to obtain the session keys. However, the suggested approach makes it difficult for the adversary to extract information from transient data because nearly all data is hashed and concatenated with the private key. This attack is sometimes referred to as Ephemeral Secret Leakage (ESL) attack by some researchers. The design of the protocol aims to thwart such attacks by securing the ephemeral session secrets and incorporating robust hashing mechanisms.

- **Privileged Insider Attack (PIA):** The system may encounter risks from both external and internal adversaries. An internal adversary with privileged access to the Trusted Authority (TA) poses a specific concern. In this scenario, the adversary may gain information about U_2 and D_3, but it's important to note that $U_2 = h(U_1||U_i)$ and $D_3 = g_3^{\frac{1}{2a+t}}$. These values are calculated using properties of elliptic curves, making it computationally difficult to reveal U_2 and D_3. Additionally, even if the intruder manages to fetch other information from the data carrier, this task is challenging due to the security measures discussed in the preceding properties of the protocol. The design of the protocol aims to withstand potential attacks, even from insiders with privileged access.

- **Fog-node Impersonation Attack (P-IA):** In the authentication phase, an adversary might attempt to input a value U_i to impersonate a user or fog node U. However, completing the authentication process by the Trusted Authority (TA) requires U_2^*, which is concatenated and hashed with other parameters by the TA. This design makes it computationally infeasible for an adversary to successfully impersonate a fog node U, as the adversary would lack the required authenticating parameters for the TA. The protocol's authentication mechanism adds an extra layer of security against impersonation attempts.

- **User Impersonation Attack (U-IA):** Even if an adversary manages to obtain the identity of a user, denoted as d_i, attempting to impersonate the user in the authentication process with the Trusted Authority (TA) poses a significant challenge. The authentication involves the calculation of D_3 and SK_{DT}. Importantly, D_3 is computed using the generator of the elliptic curve, making it computationally infeasible for the adversary to ascertain the required parameters needed to successfully impersonate the user. The protocol's reliance on elliptic curve properties enhances the security of the authentication process, preventing unauthorized users from impersonating legitimate ones.
- **Trusted Authority (TA) Impersonation Attack (TA-IA):** In attempting to impersonate the Trusted Authority (TA) and acquire the key SK_{TA}, an adversary would need to calculate T_u and T_D. These values are defined as $T_u = h(B_1 |B_2||U^1||T^1)$ and $T_D = h(D_1||D_2||D^1||T^1)$, where the terms B_1, B_2, D_1, and D_2 are calculated based on the generator of Elliptic Curve Cryptography (ECC). The security property of ECC, known as the Elliptic Curve Discrete Logarithm Problem (ECDLP), makes it exceedingly difficult to determine these terms. Consequently, attempting a Trusted Authority Impersonation Attack (TA-IA) is not practically feasible within this protocol. The use of ECC properties adds a strong layer of security, preventing adversaries from impersonating the Trusted Authority.
- **Known-key Attack (K-KA):** In the proposed protocol, the compromise of one session key does not imply the compromise of all session keys. The approved session key is derived from three distinct, session-specific ephemeral secrets, namely T_U and T_D. Due to the complexity of the Elliptic Curve Discrete Logarithm Problem (ECDLP), which is an inherent property of Elliptic Curve Cryptography (ECC), it is possible that the adversary may not simultaneously derive all of these ephemeral secrets. As a result, if one session key is disclosed, the adversary's ability to discover other session keys is restricted. This property is often referred to as the No Key Control property by researchers, highlighting the protocol's ability to limit the impact of a compromised session key on the overall security of the system
- **Replay Attack (RA):** The use of timestamps, including TS_u, TS_{TA}, and the accepted delay (δT), in the authentication phase of the protocol serves

as a safeguard against replay attacks. Replay attacks involve the malicious retransmission of previously recorded messages to gain unauthorized access or disrupt the system. In this case, the inclusion of timestamps and a predefined time window for accepted delays ensures that replaying previous messages becomes ineffective. The protocol's resistance to replay attacks is a result of this temporal validation mechanism, which enhances the overall security of the system.

- **Message Modification Attack:** In the context of a message modification attack, which involves altering an anonymous message to achieve unauthorized consequences, the proposed protocol incorporates measures to thwart such attempts.

  Firstly, the fog-node's data carrier generates U^1, which is utilized in calculating the session key. This choice of utilizing U^1 in the session key computation acts as a countermeasure against message modification. Additionally, the data carrier performs $e(B_1 \times B_2, U_1) = e(g_1, g_2)$ to verify the correctness of the user. This step further prevents message modification by ensuring the integrity of the user's identity. Moreover, the confidentiality aspect of the proposed protocol, possibly achieved through encryption or similar techniques, adds an extra layer of protection against message modification. The combination of these measures makes it challenging for adversaries to modify messages and ensures the integrity of the communication. Consequently, the proposed scheme is designed to defend against message modification attacks..

- **Resistance to Bogus Message Attack:** In the proposed scheme, a fake message would fail the message correctness test as every fog-node and user independently verifies the accuracy of the received message against established conditions. Each user or fog-node checks the validity of the message, and if it does not meet the required criteria, it will be discarded. Consequently, fraudulent message attacks are unlikely to succeed against the scheme. The message correctness test acts as a robust defense mechanism, ensuring that only valid and properly authenticated messages are accepted, thereby enhancing the overall security of the system.

- **Unlinkability:** Each time, the Anonymous Attestation Certificates (AACs) are transmitted by the fog-nodes or users they are distinct. This uniqueness is attributed to the short lifespan of session keys, which are arbitrary random numbers used in generating these certificates. The anonymous certificates and signatures lack any correlation due to the variability of these random integers. Consequently, an attacker is unable to establish links between multiple anonymous certificates and signatures generated by the same fog-node. This characteristic renders our scheme unlinkable result.

- **MVAM Attack:** The primary processing component in our system is the Trusted Authority (TA). It is important to acknowledge the potential vulnerability to memory-related attacks, like Multi Variant Attack on Memory (MVAM) or Direct Memory Access (DMA) attacks, which could target the TA [218]. Our scheme uses SK_U and Sk_{TA}. If these are equal then DC authenticates the trusted authority, otherwise not.
- **DDoS Attack:** Adversaries might employ numerous compromised or zombified sensors/IoT devices to launch a Distributed Denial of Service (DDoS) attack against a single server. In our proposed protocol, each user and IoT device undergoes initial registration with a unique key, receiving a verification key for each specific session in return. The Trusted Authority (TA) then verifies this key in every session, effectively preventing non-legitimate users/devices from gaining access to that particular session.
- **MITM Attack:** In the proposed protocol, mutual authentication occurs between the user, fog-node, and Trusted Authority (TA). If an adversary attempts a Man-in-the-Middle (MITM) attack to intercept communication between the user, fog-node, or TA, they would need to create fake beacons. However, the BDHP (Bilinear Diffie-Hellman Problem) asserts that gaining access to information is highly challenging. Therefore, our protocol renders the MITM attack ineffective by making it difficult for an adversary to penetrate the exchanged information.

8.6 Conclusion

In today's world, IoT systems play a vital role and are essential for the functioning of various aspects of life. Ensuring the security of IoT is a critical concern to safeguard humanity from potential disasters. This study focused on developing a robust authentication scheme resilient to attacks. The research thoroughly examined various attacks and assessed the proposed scheme's resilience. While specifically designed for Fog-centric IoT systems, the proposed solution can be extended to enhance the security of diverse systems, including the Internet of Medical Things (IoMT), drones, and smart grids.

Chapter 9

Navigating Ethical Challenges: Frameworks for Swarm Intelligence in Fog Computing

*Wasim Khan** and *Mohammad Ishrat*

In an era defined by rapid technological progress, the fusion of swarm intelligence and fog computing stands as a crucial juncture, offering transformative possibilities across various sectors. This chapter provides a comprehensive examination of these two emerging technologies, focusing on their technical underpinnings, ethical implications, and practical applications. Swarm intelligence, drawing inspiration from natural phenomena, showcases decentralized control and collective behavior. This approach marks a departure from traditional centralized systems, paving the way for more adaptive, scalable, and robust technological solutions. Fog computing emerges as a key player in the era of big data and the Internet of Things (IoT), extending cloud computing to the network's edge. This shift significantly enhances the efficiency of data processing and decision-making, reducing latency and bringing computation closer to data sources. The convergence of swarm intelligence and fog computing presents a potent com-

Department of CSE, Koneru Lakshmaiah Education Foundation, Vaddeswaram, Andhra Pradesh, India.

* Corresponding author: khanwasim051@gmail.com

bination that promises to revolutionize various domains, including agriculture, smart cities, and healthcare. This integration leads to systems that are not only more efficient and responsive but also capable of tackling complex tasks that were previously unmanageable. However, the advancement of these technologies brings forth significant ethical challenges. The chapter delves into issues such as privacy, security, and accountability, highlighting the need for comprehensive ethical frameworks and guidelines. It emphasizes the importance of developing these frameworks to guide the responsible deployment of swarm intelligence and fog computing, ensuring that they align with societal values and norms. The chapter also explores the technical foundations of both fields, detailing the principles of swarm intelligence like decentralization and self-organization and the layered architecture of fog computing. It discusses the integration of these technologies and their application in real-world scenarios, demonstrating their potential to create efficient, scalable, and resilient systems. In conclusion, this chapter calls for continued research and the practical application of swarm intelligence and fog computing, advocating for a balanced approach that integrates technological innovation with ethical considerations. It underscores the imperative for collaborative efforts among researchers, practitioners, and policymakers to guide these advancements towards a future that is innovative, ethically sound, and socially beneficial.

9.1 Introduction

Swarm intelligence, a cutting-edge field that draws inspiration from the group behavior of natural systems like ant colonies, bird flocking, and fish schooling, has emerged as a crucial strategy for resolving complex, decentralized problems. This approach is particularly effective in optimization and decision-making processes and finds significant applications in the realm of fog computing.

Particle Swarm Optimization (PSO) is a key method in swarm intelligence inspired by how birds and fish move together. It is efficient at solving tricky problems by having each 'particle' (like a bird or fish) move around to find the best answer. They learn from what they've done before and what their friends have done. This way, they all eventually find the best solution together [80].

Ant Colony Optimisation (ACO) is another important method, and it's all about how ants find food. It's great for figuring out the best paths, such as planning the quickest delivery routes or making schedules. In ACO, the ants in the computer leave trails like real ants do with scent, helping find the shortest way [78].

Bee Colony Optimisation (BCO) gets its ideas from how bees find flowers. Bees do a dance to tell other bees where to find good flowers. BCO uses this idea to solve big, complex problems. Each 'bee' in the program checks out different solutions and shares what it finds, helping to choose the best one [148].

The Firefly Algorithm (FA) is inspired by how fireflies light up to attract each other. This method is used to find good solutions in very complicated situations. Each 'firefly' in our program shines brighter for better solutions, and the others follow the brightest one [115].

Bacterial Foraging Optimization (BFO), is the final method, which is based on how bacteria move towards food. This method is used in things like designing or planning, where finding the best solution is like bacteria finding the best food. These methods are extremely useful in fog computing. Fog computing is like a big network without one main boss, and these methods help manage resources, balance work, and handle lots of data. They fit really well with fog computing because they work on their own and adapt to changes [94].

Research in swarm intelligence and fog computing shows how important it is to combine the two fields in this era of rapid growth [43]. This chapter talks about the basics of fog computing and swarm intelligence. It also talks about their moral effects and how they can be used in the real world. Swarm smartly models making decisions and working together in real time based on what is happening at the moment. The way decisions were made before was controlled, but from now on, they will be self-managed. The people who work in technology have done a great job of using this idea to make a complicated system that can change and work like wild species do. Swarm intelligence is changing the way we solve hard problems and how we do transportation and search and rescue (SAR) [62]. The Internet of Things, cloud computers, and big data have all helped it gain a reputation as a useful tool. By adding the network's edge to the idea of cloud computing, the goal is to move data processing and analysis closer to where the data comes from. Architectural flexibility helps data handling run more smoothly and quickly, which is great for real-time apps that need to make decisions and run quickly [87, 133].

When swarm intelligence and fog computing work together, they can help with smart cities, healthcare, farming, and emergency management, among other things [161]. When different technologies are combined, machine performance will go up, letting them do previously unimaginable, complicated jobs. Putting these two jobs together could lead to technological progress. We need strong, efficient, and scalable fog computing technology based on swarm intelligence. This section on working together across agencies talks about a lot of different topics, from problems with technology to moral and practical issues. By having a lot of power there is an expectation that you have the responsibility to take care of things. As technology advances, it raises the following social questions.

1. How can these processes be more open and look to the future?
2. How would these changes make people feel about their safety, privacy, and traditions?

This section goes into these problems in more depth, as well as the moral issues that swarm intelligence and cloud computing bring up. This piece talks about how ethical principles and good design must be taken into account when creating these technologies in a way that improves people's health [182].

As fog computing and swarm intelligence become more common, new problems and moral questions come up. We need to carefully look at new technology in relation to societal models and goals if we want to find answers to these issues. Cloud computing and AI are interesting new fields of study, and this article lays the groundwork for more research to come. This section gives readers a better idea of these technologies by looking at where they came from, how they should be used, and what their ethical issues are. It also sets the stage for the rest of the chapter. This chapter gives you a basic rundown of the information you need to stay current in crowd intelligence and cloud areas, which are always changing. This article's goal is to help readers understand these ideas and the problems and chances they bring to the field of justice science and technology so that they can make smart choices.

9.2 Technical Foundations

Some of the main ideas and building blocks of swarm intelligence and fog computing are different, but they work well together. You should learn more about these building blocks if you want to know how they might change systems that process data and make decisions.

9.2.1 Swarm Intelligence Principals

A few basic ideas form the basis of swarm intelligence: Decentralization has prevented a single power from controlling the actions of various actors. In fact, swarm bots follow a set of rules that have already been decided upon and have their own unique point of view. The system's architecture is strong enough to handle the failure of a single agent without affecting the system's general performance because it is distributed [173].

- One thing that makes swarm systems unique is their ability to organize themselves without outside help. At the local level, when agents connect with their surroundings, self-organization starts to happen. From there, world patterns or behaviors that are common to everyone start to show up [99].
- In swarm intelligence, multiple agents work on tasks that overlap or are similar, which creates repetition. Because some agents can't work or are blocked by this redundancy, the system is strong even when something goes wrong or the world changes [84].

9.2.2 Fog Computing and Architecture

Fog computing is a powerful technology that brings the advantages of cloud computing closer to where data is generated and used. Imagine a layer of computational and storage resources distributed over a network. This layer sits between the end devices (like IoT devices) and the traditional cloud, offering quicker data processing and reduced latency. Here are some key characteristics of fog computing:

- Fog nodes, which are small-scale data centers or devices capable of processing and storing data, are positioned close to where data is generated. This proximity significantly reduces the distance data has to travel compared to traditional cloud computing.
- The reduced distance leads to a notable decrease in latency. This means applications that rely on fog computing respond much quicker, which is crucial in scenarios like emergency response systems or real-time financial trading.
- Fog computing enables the processing of data almost as soon as it's generated. This is essential for applications that require immediate insights from data, such as monitoring vital signs in healthcare or detecting anomalies in industrial processes.
- By enabling local data processing, fog computing allows for decisions to be made quickly at the network edge without the need for data to be sent to a distant cloud server.
- Fog computing infrastructure can be scaled up or down, depending on the workload. This flexibility is advantageous for businesses experiencing fluctuating data processing demands.
- As businesses grow and their data requirements increase, fog computing environments can expand to accommodate this growth, supporting a wide range of devices and applications.
- By processing data locally, fog computing minimizes the amount of sensitive data travelling over the network, reducing the exposure to potential breaches.
- It allows for more robust security protocols at the local level. Each fog node can be equipped with its own security measures, making the system as a whole more resilient against cyber threats.

In layered fog computer architectures, the following parts are often found:

- The gadgets that send or receive data are on the edges of a network. Things like cameras, sensors, and Internet of Things devices demonstrate this.
- Near the edge are computing devices such as switches, routers, and other fog nodes. The cloud takes care of initial processing, analytics, and storing, so not all data needs to be sent there.
- Fog servers perform cloud-based services, such as local processing. The cloud is responsible for complex processing and long-term data storage. With this layered system, which uses both on-premises and cloud tools, it is easier to process, store, and handle data.

9.2.3 Swarm Intelligence Integration with Fog Computing

- Swarm intelligence integration with fog computing can make the system much more useful.
- The ideas behind swarm intelligence make the best use of resources and keep nodes from getting too busy by assigning computing jobs to fog nodes in the best way possible.
- In a fog computing setting, swarm algorithms can improve the speed and efficiency of data transfer.

9.2.4 Real World Use Cases

Swarm intelligence has been successfully used in fog computing environments in a number of case studies and real-world applications. Here are some examples:

- *Smart Traffic Management:* The use of fog computing to process data from many sensors and cameras around a city in real time with swarm intelligence algorithms helps improve traffic flow in smart city apps. This helps with smart traffic control.
- *Systems for keeping an eye on healthcare:* Wearable medical devices send data to fog computing nodes that use swarm intelligence algorithms. These nodes can then process the data and send real-time health tracking and alarms.
- *Drones for farming:* In fog computing settings, processing data and making decisions happen at the edge so that responses happen right away. Swarm intelligence is used to coordinate agricultural drone fleets for jobs like keeping an eye on crops and getting rid of pests.

9.3 Ethical Implications of Integrating Swarm Intelligence in Fog Computing

Putting swarm intelligence and fog computing together brings up a lot of moral questions, like those about privacy, fairness, discrimination, and safety as given in Figure 9.1. These are tough topics that cover a lot of important ground.

Security Issues: Safety is also very important. Swarm intelligence nodes are used in fog computing systems. These nodes are open to many security risks because they are spread out and often act on their own [72, 142]. Imagine that hackers broke into the swarm network and took over the system for distributing energy after a smart grid event. A security breach at just one node could put the whole system at risk because all the gadgets in the network depend on each other. Cyberattacks are impossible if these systems aren't kept up-to-date and protected by strict security rules.

The Swarm Intelligence products from Fog Computing are poorly designed, which makes security worse [25]. For instance, think about a business setting where an attacker could possibly take over the whole network with just one IoT device that they got into. In the case of industrial automation, a broken monitor caused the production line to shut down, which cost a lot of money. These attacks can be less damaging if you have security measures in place, such as real-time monitoring and the ability to spot strange behavior. You should also think about how to build security into the system from the very beginning.

Decision-Making Accountability: It's challenging to determine how swarm intelligence powers decision-making in a fog. It can be hard to figure out who is to blame when actions have bad or unexpected results, even if the systems can make decisions on their own thanks to joint algorithms. When computers make decisions that affect many people, like in healthcare or self-driving cars, it raises a lot of difficult moral and legal questions. A clear set of accountability standards is needed to make sure that everyone who is responsible for making, managing, or using the technology knows what they need to do [233, 133].

Figure 9.1: Ethical implications of integrating Swarm Intelligence in Fog computing.

Because swarm intelligence systems make choices both on their own and with other systems, it is harder to figure out who is at fault. A network of linked cars and traffic nodes makes decisions in real time in autonomous traffic management systems. This means that one wrong move could have disastrous effects. Sometimes, inputs and strategies from many nodes get mixed up in a messy way during the decision-making process, making it hard to find the starting node. Rules that haven't kept up with how quickly technology changes have made it harder to know who is responsible for what.

Data Privacy Issues: There will always be privacy issues when fog computing and swarm intelligence are used together because they make the network for processing and sharing data very complicated. The systems in these places handle and store huge amounts of data, some of which is private and sensitive. A smart city app case study showed that there are big privacy risks, even though phone data can help plan cities. Users' safety could be at risk if this data is misused or viewed without permission. Data is not stored on a single server in swarm intelligence. Instead, it is spread out among many nodes. This makes things even stranger [137].

Data privacy is more important in fog computing because swarm intelligence processes and saves a lot of different kinds of data across many nodes. One example is how healthcare apps send data about patients over a network so that it can be processed and analyzed. It is possible to gather different kinds of patient data, such as basic health records and very private genetic information. There are many risks that come with sharing information. In a notable case study involving a smart healthcare system, patient data was accidentally shared because the swarm network did not have enough security measures in place. This caused a big breach of privacy. To fix these issues, we need strong security measures, open data governance standards, and data processing tools that let users see and control their own data.

Bias and Fairness: Unintentional bias is one thing that could go wrong with swarm intelligence systems. This makes the training set or program skewed most of the time. Two places where this could lead to unfair results are the hiring process and planning police shifts ahead of time. One thing that could lead to unfair hiring practices is using platforms that use skewed algorithms to treat some groups of people better than others. It's necessary to do regular audits, explain algorithmic methods, and use different datasets in order to get rid of bias and get equal results [220].

Swarm intelligence systems make it clear that biased or unfair decisions made by computers on their own are a problem. Because of how complicatedly their many algorithms work together, these systems tend to show emergent rather than set traits. Because credit score systems use biased data or algorithms, some groups of people may be unfairly turned down for loans or have their interest rates raised.

Because inequality might be against the law, this makes us consider ethics and the law. To get around these problems, you need to use different sets of data, algorithms that are aware of justice, and keep checking the results for skewing. Accountability and openness in algorithms go hand in hand. It is important for algorithms to be open so that people can be trusted and for people to be responsible.

Impact on Economy and Society: Fog computing systems built on swarm intelligence are bad for society and the economy because they use a lot of energy and resources. These systems might use a lot of energy because they need to constantly handle data and talk to each other. These technologies not only make people more aware of how important it is to use renewable and long-lasting energy sources, but they also make people worry about how these technologies might affect the world. There are environmental issues with all of the gadgets that these networks use. Concerns have been raised about how to properly handle and recycle electronic trash that comes from making, using, and finally getting rid of electronic components. These activities produce a lot of electronic waste. So, it's more important than ever to find ways to make these technologies less harmful to the world without making them less useful [163, 134].

Sociologists say that the widespread use of swarm intelligence in fog computing could have huge effects on many parts of society. The digital gap could get even bigger, which is a big worry. Some groups may start to enjoy a higher level of technological luxury, especially in places with more advanced technology. Because of this, the gap between those who have access to good technology and the rest of society may grow. This split could have a lot of different, far-reaching effects. There are differences in two of them: the number of economic chances and the basic services that are available. Another major concern is that these technologies might result in more job losses as a result of the increased use of machines to complete tasks. A lot of people are afraid that technology will cause them to lose their jobs in fields where it is already common. After this change, people will have to think about their jobs again, learn new skills, and maybe even start over if they want to work in a world where AI is more essential. Making sure that everyone who is affected by new technology gets fair help and rewards is the right thing to do.

Explainability and Transparency: Because these technologies are so hard to understand and use, the ideas of explainability and openness in swarm intelligence in fog computing are becoming more popular [112]. This is what we mean when we talk about transparency: how open and clear the ways of making decisions and handling data are. Due to Swarm Intelligence's decentralized and independent decision-making process, it might be hard to track down and understand some results. This lack of clarity is especially annoying in fields like healthcare, law, and finance, where it's important to know who makes the choices.

"Explainability" refers to how easy it is for people to understand how a system makes choices. A lot more work needs to be done when a group of nodes, which are also called networked bots, work together to make decisions. People have a hard time predicting or explaining how the system will act because it is so connected and acts in strange ways all the time. A swarm intelligence-based traffic control system has a lot of moving parts and no central decision-maker. This means that it might be hard to explain why certain routing choices were made during rush hour [107]. Being honest and responsible makes people want ethical standards that are clear and easy to understand. People who use and depend on these systems must not only be able to believe that they are true and accurate, but they must also be able to hold the system and the people who made it responsible for what it does. And this is even more important when these systems make choices that have a big effect on people's lives. More and more people want XAI technology to be developed because they need to understand these complicated systems and find ways to fix them. Making systems that can act quickly and explain what they do is the only way to achieve this goal. Using AI and CS ideas and principles to better understand and describe things is becoming more and more strictly required by law in more and more places.

9.4 Ethical Frameworks in Emerging Technologies

In the realm of artificial intelligence (AI) and robotics, several principal ethical frameworks have been established to guide developers and users towards responsible and beneficial usage. These frameworks play a crucial role in shaping the development and implementation of AI technologies.

9.4.1 Overview of Existing Ethical Frameworks in AI and Computing

In the rapidly evolving landscape of AI and computing, ethical considerations have become increasingly paramount [234, 135]. The swift advancement of AI technologies has introduced complex ethical dilemmas, necessitating the development of comprehensive frameworks to guide ethical AI development and deployment. This section aims to explore these frameworks shown in Figure 9.2, emphasizing their relevance and adaptability to the fields of AI, including specific applications like swarm intelligence and fog computing.

The Asilomar AI Principles, established during a 2017 conference by leading AI researchers, stand out as a cornerstone in this area [176]. These principles are centered around research issues, ethics and values, and long-term concerns. They advocate for the development of AI in a way that benefits all of humanity, emphasizing the need to prioritize human values and avoid a race to the bottom in AI development standards. The principles underline the importance of cooper-

Figure 9.2: Ethical frameworks.

ative and beneficial AI, ensuring that these technologies, while powerful, remain aligned with human values and ethical considerations.

Another pivotal framework is the IEEE's Ethically Aligned Design. This comprehensive set of recommendations offers a deep dive into how AI and autonomous systems can be developed while prioritizing human rights, well-being, data agency, and effectiveness. It presents a detailed approach to ensuring that AI systems respect and enhance human dignity, freedom, and cultural diversity. This framework is particularly significant in its emphasis on ethical, transparent, and accountable system design, aligning technological advancements with human-centered values [11]. The European Union's Ethics Guidelines for Trustworthy AI also play a critical role in shaping the ethical landscape of AI. These guidelines, created by the High-Level Expert Group on Artificial Intelligence, put a strong emphasis on ensuring the reliability of AI systems. They are founded on key principles such as transparency, justice and fairness, non-maleficence, responsibility, and privacy. These guidelines encourage the development of AI systems that are not only technologically robust but also ethically sound and respectful of human rights, democratic values, and the rule of law [229].

While these frameworks vary in their specific focus and recommendations, they collectively provide foundational guidelines that can be applied across the spectrum of AI technologies. Their broad principles are adaptable to specific AI applications, making them particularly relevant to emerging fields such as swarm intelligence and fog computing. With their own set of ethical issues and traits, these areas can really benefit from using these well-established ethical frameworks. This will make sure that their growth is in line with the main goal of helping everyone while still respecting individual rights and societal values.

9.4.2 Adapting and Developing Ethical Frameworks for Swarm Intelligence in Fog Computing

Adapting and developing ethical frameworks for the specific realms of swarm intelligence and fog computing is crucial, given their unique characteristics and the ethical challenges they present. This section aims to explore how the general principles of existing ethical AI frameworks can be fine-tuned to these emerging technologies [198].

Swarm intelligence, characterized by its decentralized and autonomous nature, raises unique challenges in terms of accountability and transparency. In swarm-based systems, decision-making processes are distributed across numerous agents, each operating based on local rules and interactions rather than a centralized command. This decentralization, while advantageous for scalability and resilience, complicates the traceability and accountability of decisions and actions [99]. To address these challenges, existing ethical frameworks should be adapted to include guidelines for distributed decision-making. These adaptations could involve developing standards for documenting and reporting the decision-making processes within swarm systems, ensuring that each agent's role and the collective behavior can be understood and assessed. Emergent behavior is a feature of swarm intelligence. It happens when simple local interactions lead to complex global patterns. This means that there needs to be ethical rules that can predict and control these outcomes, making sure they are in line with human values and morals.

Fog computing, which extends cloud computing to the edge of the network, presents its own set of ethical challenges, especially regarding data privacy and security [47, 204]. The distributed nature of fog computing means that data is processed and stored closer to where it's generated, which can enhance efficiency and reduce latency. However, this also raises concerns about the security and privacy of the data at these edge nodes. Adapting ethical frameworks to fog computing would involve emphasizing data governance principles that are specific to the distributed architecture. This includes ensuring that data is processed and stored securely at the edge, implementing robust encryption methods, and establishing clear protocols for data sharing and transfer. It's also crucial to develop guidelines for local data processing, ensuring that personal and sensitive data is handled appropriately, respecting user privacy, and complying with relevant data protection regulations. There is also a pressing need to develop new ethical guidelines or frameworks that specifically address the blend of swarm intelligence with fog computing [88]. These technologies, when integrated, could lead to novel applications and systems with capabilities and ethical implications that are not fully covered by existing frameworks. For instance, the combination of swarm intelligence's decentralized decision-making with fog computing's distributed data processing could result in complex systems where accountability and data governance become intricately interwoven. New ethical guidelines should consider these synergies and the potential for unforeseen ethical dilemmas, offering a structured approach to evaluate and guide the development and deployment of these integrated systems.

In summary, adapting existing ethical frameworks and developing new ones for swarm intelligence and fog computing is essential to ensuring that these technologies evolve in a way that is not only innovative but also ethically responsible and aligned with societal values. These adaptations and new developments should address the unique challenges posed by the decentralized, autonomous nature of these technologies, ensuring transparency, accountability, data privacy, and security.

9.4.3 Role of Regulatory Bodies and Standards

The role of regulatory bodies and international standards organizations is pivotal in shaping ethical AI practices, including those in specialized areas like swarm intelligence and fog computing [117, 132]. This section aims to discuss the influence of such bodies and standards, their impact on AI technologies, and their specific implications for swarm intelligence and fog computing.

Regulatory bodies like the European Commission and standards organizations such as the International Organization for Standardization (ISO) play a critical role in influencing the development and ethical alignment of AI technologies. They do this by establishing regulations and guidelines that set the benchmark for responsible AI development and deployment. These regulations often focus on ensuring that AI systems are developed in a way that respects human rights, privacy, and ethical norms. Standards organizations, on the other hand, provide guidelines and best practices that help in the standardization of AI technologies, ensuring consistency, safety, and quality across different applications and industries [145]. One of the most notable examples of such regulation is the General Data Protection Regulation (GDPR) in Europe. GDPR has set a high standard for data privacy and user consent worldwide, impacting how AI systems handle personal data. For fog computing environments, where data processing and storage occur closer to the data source (and hence, the user), GDPR has significant implications. It mandates stringent data protection and privacy measures, requiring that systems built on fog computing architectures adhere to principles like data minimization, purpose limitation, and user consent. These requirements ensure that personal data is not only processed securely but also that users have control over their data, a crucial aspect in building trust in these technologies [27].

Looking to the future, it's likely that we will see more targeted regulatory developments focusing on specific aspects of swarm intelligence and fog computing. Given the decentralized nature of swarm intelligence, regulations might evolve to address issues of accountability and transparency in systems where decision-making is distributed across multiple autonomous agents. For fog computing, future regulations might focus more on data security and privacy at the edge of the network, setting standards for how data is processed, stored, and shared in these distributed environments [143]. It's crucial for technologists and policymakers to proactively engage in shaping these regulations. This engagement is necessary to ensure that the regulations are not only feasible but also foster innovation while upholding ethical standards. Regulations that fail to recognize the distinctive qualities and potential of these technologies shouldn't impede their development. Similarly, ethical compliance should not be an afterthought but a core aspect of the development process. This proactive engagement will help strike a balance between innovation and ethical compliance, ensuring that advancements in swarm intelligence and fog computing are both ethically responsible and technologically progressive.

9.5 Implementing Ethical Practices

In this section, a detailed exploration of practical approaches to embedding ethical considerations in swarm intelligence and fog computing technologies is presented. This exploration is crucial for ensuring that the development and application of these technologies are aligned with ethical standards and societal values.

9.5.1 Strategies for Ethical Data Management

In the context of fog computing, where data is often processed at the network's edge, close to the data source, the importance of robust data security measures is heightened [202]. Advanced encryption techniques become crucial in this scenario, as they protect data during transmission and storage, safeguarding it from potential breaches and unauthorized access. This encryption ensures that data, even if intercepted, remains unintelligible and secure. Strict access controls form another pillar of ethical data management. These controls ensure that data access is restricted to authorized personnel or systems, significantly reducing the risk of unauthorized data manipulation or breaches. This level of control is crucial, especially in decentralized systems like those in fog computing, where data is processed and stored across multiple nodes.

Furthermore, transparent data sharing protocols are essential in maintaining the ethical integrity of these systems [202]. In an era where data privacy concerns are paramount, it is important for systems to have clear and transparent policies regarding the usage and sharing of data. Central to these policies is the principle of user consent and control. Users must have clear information about how their data is being used and should be able to control the sharing of their data. According to laws like the General Data Protection Regulation (GDPR) in Europe, following user consent is not only consistent with ethical data management practices but also a legal requirement in many jurisdictions. Compliance with such regulations is not merely a legal obligation but also a critical factor in maintaining and building public trust in these technologies.

In summary, the adoption of strategies for ethical data management is essential in the realms of swarm intelligence and fog computing. These strategies, encompassing data minimization, robust data security measures, strict access controls, and transparent data sharing protocols, ensure that the handling of data in these complex systems is carried out ethically, responsibly, and in compliance with legal standards. This approach is not only crucial for safeguarding user privacy and system security but also for maintaining the integrity and trustworthiness of these advanced technological systems.

9.5.2 Ensuring Transparency and Accountability in Algorithms

These systems, characterized by their complexity and often opaque nature, require a high level of transparency in their algorithmic decision-making processes. Transparency is essential not only for the users and stakeholders to understand and trust the system but also for ensuring the system operates within ethical boundaries.

Transparency in algorithmic decision-making in swarm intelligence systems involves clearly explaining how decisions are formulated and executed by the system [57]. This clarity is vital for fostering trust among users, as it allows them to understand the rationale behind specific actions or outcomes. Moreover, transparent systems enable stakeholders to identify and rectify potential biases or errors in the algorithms. In an era where algorithmic decisions can have significant impacts, the ability to scrutinize and understand these decisions is crucial. However, achieving transparency in swarm intelligence systems, which are inherently complex due to their decentralized and autonomous nature, presents unique challenges. Traditional centralized systems may have a clear decision-making hierarchy, but in swarm intelligence, decisions emerge from the interactions of numerous autonomous agents. Therefore, it's important to develop methodologies that can trace and explain these decision-making processes. This might involve creating systems that can log and report the activities of individual agents and their contributions to collective decisions, providing insights into the workings of the swarm [212].

Accountability in these decentralized systems is another significant challenge. In a system where decisions are not made by a single entity but rather are the result of collective interactions, establishing clear guidelines and responsibilities for outcomes becomes complicated. This section addresses this challenge by proposing the establishment of clear guidelines and responsibilities, even in a decentralized setup. This involves defining the roles and responsibilities of different agents and ensuring that the overall system design aligns with ethical standards. Furthermore, the role of audit mechanisms and oversight processes is emphasized in ensuring that algorithmic decisions are ethically sound and comply with regulatory requirements. Regular audits and reviews of the system's decision-making processes can help ensure that they remain unbiased and aligned with ethical and regulatory standards. These mechanisms are crucial for maintaining the integrity of the system and ensuring that it continues to function as intended, respecting ethical guidelines and societal values. Ensuring transparency and accountability in algorithms is paramount in swarm intelligence systems. It involves making the decision-making process understandable to users, developing methods to trace decisions back to specific algorithmic components, and establishing robust audit and oversight mechanisms. These steps are crucial for building trust in these systems and ensuring they operate ethically and responsibly.

9.5.3 Ethical Design Principles for Swarm Intelligence Systems

The development of ethical frameworks and principles in the context of technologies like swarm intelligence and fog computing is essential to guiding their responsible use. These frameworks serve as the cornerstone for ensuring that the deployment and operation of these complex systems align with societal values and norms, emphasizing the importance of ethical considerations in every stage of technological development and implementation.

Development of Ethical Guidelines: The guidelines for swarm intelligence in fog computing must address the nuanced aspects of these technologies. This includes ensuring data integrity and accuracy, managing the decentralized and autonomous nature of decision-making, and addressing the potential for emergent behavior that is not immediately predictable or explainable [53]. The guidelines should cover not only the technical and functional aspects but also the societal impact, user privacy, data security, and fairness. For instance, a guideline might stipulate rigorous testing and validation procedures to identify and mitigate unintended biases in decision-making processes. Another might focus on ensuring that data collected through fog computing networks is used ethically and transparently, with clear benefits to users and minimal privacy intrusion.

Principles of Openness, Responsibility, and User Consent: These principles form the cornerstone of the proposed ethical framework. Openness in this context means transparency about how swarm intelligence systems operate, how data is processed and used, and how decisions are made. This principle aims to build trust and understanding among users and stakeholders. Responsibility involves clearly defining who is accountable for the decisions made by autonomous systems and ensuring there are mechanisms in place to address any negative consequences [219]. User consent is fundamental, emphasizing the need for users' informed consent regarding how their data is collected and used. When compared to existing frameworks in traditional IT or AI, these principles extend the focus from individual privacy and data security to encompass broader aspects of autonomy, collective decision-making, and emergent behavior typical of swarm intelligence systems.

Challenges in Implementation: Implementing these ethical frameworks is not without challenges. One major hurdle is the pace of technological advancement, which often outstrips the development of ethical guidelines and regulatory frameworks. Another challenge is the inherent complexity of these systems. The decentralized, autonomous nature of swarm intelligence, combined with the vast and dynamic nature of fog computing networks, makes it difficult to predict all potential ethical issues and address them preemptively. Furthermore, global disparities in technological development and varied cultural and legal standards across different regions add layers of complexity to the uniform application and enforcement of these guidelines. Balancing innovation and ethical responsibility, especially in a field that is rapidly evolving and has diverse applications, requires continuous dialogue among technologists, ethicists, policymakers, and users [193].

Inclusivity and Diversity: This emphasizes the importance of considering a diverse range of perspectives and experiences in the development of these technologies. To ensure that these technologies are available and helpful to a diverse population, it is crucial that the design and implementation of swarm intelligence and fog computing systems take into account a wide range of societal needs and values. This includes addressing potential biases in data sets and algorithms to prevent discriminatory outcomes [28].

Adaptability and Flexibility: Given the rapid evolution of technology, ethical frameworks must be adaptable and flexible. It addresses the need for ethical guidelines that can evolve in response to new developments and unforeseen challenges in the fields of swarm intelligence and fog computing. It involves establishing mechanisms for regular review and update of ethical standards [88].

Human Oversight and Control: It addresses the need to maintain human oversight and control over autonomous systems. Despite the advanced capabilities of swarm intelligence in fog computing, it is essential to have mechanisms in place that allow for human intervention, ensuring that decisions made by these systems can be overridden or altered by human operators when necessary [222].

Ethical Risk Assessment: This involves the proactive identification and assessment of potential ethical risks associated with the deployment of these technologies. It calls for the implementation of comprehensive risk assessment strategies to identify and mitigate potential negative impacts on individuals and society [136].

In summary, the Ethical Design Principles for Swarm Intelligence Systems emphasize the importance of creating systems that are transparent, respect user autonomy and privacy, ensure fairness and non-discrimination in decision-making, and are adaptable to ongoing ethical challenges. These principles are essential for ensuring that swarm intelligence systems are not only technologically advanced but also ethically responsible and aligned with human values and societal expectations.

9.6 Case Studies: Ethical Dilemmas and Resolutions

Several real-world scenarios are presented in this section to illustrate the ethical challenges these fields face. This section not only provides an analysis of these challenges but also offers insights into possible resolutions, drawing lessons and best practices from each case.

9.6.1 Real-World Scenarios Illustrating Ethical Challenges

In the exploration of real-world scenarios, two pertinent cases are presented that exemplify the complex ethical landscapes in which swarm intelligence and fog computing operate.

Swarm Robotics in Disaster Relief Operations: In this scenario, the deployment of swarm robotics for disaster relief efforts is examined. Here, a fleet of drones is programmed to perform critical tasks, including delivering emergency supplies and surveying for survivors in a disaster-stricken area. The ethical challenge arises from the autonomous decision-making capabilities of these drones. For instance, if the drones encounter a situation where they must choose between continuing to search for survivors or delivering life-saving supplies to a known group of stranded individuals, the decision-making process becomes ethically complex. This scenario poses significant ethical questions: Should the drones prioritize the delivery of supplies over the search for additional survivors, or vice versa? The ethical dilemma intensifies, considering the potential life-or-death consequences of these decisions. The autonomous nature of the swarm intelligence system necessitates an ethical framework that guides these critical decision-making processes, ensuring that the drones' actions align with humanitarian principles and values.

Due to unpredictability, rescue and relief efforts are always late, and the havoc wrought can only be seen after the tragedy has reached its peak. Gladence et al. [89] dispatches a fleet of unmanned airborne vehicles (UAVs) and a swarm of airborne robots (drones) that work together to achieve a goal: to gather damage and disaster information. All swarm drones VOTL, that is they, vertically hover, take off, and land. Every drone has a certain number of propellers. Decentralized, natural, or artificially self-organized systems exhibit swarm intelligence. This behavior relies on AI. Flooding is simulated. Drone swarms will congregate and assess catastrophe damage. In order to carry out an effective SAR operation in an unfamiliar setting, Din et al. [62] introduced a behavior-based model for a swarm of tiny robots. An algorithm chooses a virtual leader (VL) at runtime to direct the swarm. The search targets are victims, and the process is self-contained and dynamic, based on a behavior-based approach. As the swarm travels from its current location to the referred targets under the suggested paradigm, the VL keeps it intact. Ganesan et al. describes miniature disaster-relief robots that use swarm intelligence routing to detect an object or person within a building and determine the quickest way outside [81]. Many tiny robots work together to achieve redundancy and remote access. Each robot uses cameras, ultrasonic sensors, temperature sensors, and more to find objects in varied lighting situations. Communication, dynamic isolation of failing robots, and network formation are possible with these robots. Ant colony optimization (ACO) methods are also discussed in this work to discover the shortest route between the object of interest and the outside.

Fog Computing in Healthcare: The second scenario delves into the application of fog computing in the healthcare sector. In this setting, patient data is processed at multiple points in a distributed network, which can enhance the efficiency and responsiveness of healthcare services. However, this decentralization poses significant ethical challenges, particularly concerning patient confidentiality and data security. As patient data traverses various nodes in the network, ensuring its confidentiality becomes complex. The ethical challenge here is to maintain stringent data protection measures across all points of the network while reaping the benefits of fog computing. This scenario demands robust encryption protocols, strict access controls, and comprehensive data governance policies to safeguard sensitive patient information. The ethical implications extend to ensuring that patients' rights to privacy are respected, balancing the technological advantages of fog computing with the imperative of maintaining trust in healthcare services. Elhadad et al. [68] proposed a novel fog computing-based system for healthcare monitoring with the goal of handling real-time notifications. Wearable sensors measure vital signs like heart rate, temperature, and blood pressure, and the suggested system would alert medical professionals instantly if any of these values deviate from the normal range, all with the help of machine learning algorithms. It is also possible to set up notifications to remind patients to take their meds or stick to their diet at certain intervals. For the benefit of hospitals and researchers in the future, the cloud layer saves enormous amounts of data to the cloud.

In order to efficiently gather and analyze data, Paul et al. [190] suggest using fog computing to assist in the monitoring of patients with chronic conditions in this study. The most difficult part would be identifying which context-sensitive data is actually useful for the patient's health. Adding a fog computing layer improves upon a basic sensor-to-cloud architecture, which is no longer practical. This improves the system's overall efficiency since it decreases the data transfer rate between the cloud and the sensors and gets rid of the risk associated with data center failure. They also took a look at the fog computing layer's deployment and security concerns. Roy et al. [213] provided a non-DHT-based peer-to-peer (P2P) resource discovery architecture that is based on publish/subscribe and interest/resource. Scalability, fine-grained access control to messages, and increased security are all features of the publish/subscribe communication architecture, which makes it ideal for real-time handling of massive data and message volumes.

These scenarios highlight the multifaceted nature of ethical challenges in the realms of swarm intelligence and fog computing. They demonstrate how advancements in these technologies bring forth complex ethical considerations that must be meticulously addressed to ensure these innovations contribute positively to society and do not inadvertently cause harm or ethical transgressions.

9.6.2 Analysis of Responses and Solutions

This section delves into how the ethical challenges presented in the real-world scenarios of swarm robotics in disaster relief and fog computing in healthcare can be addressed through specific solutions and strategies.

In the Disaster Relief Scenario: The deployment of swarm robotics in disaster relief presents the challenge of making critical, autonomous decisions that could significantly impact human lives. To address this, the implementation of ethical decision-making algorithms is proposed. These algorithms would be transparent and accountable, allowing third parties to understand and assess the decision-making process. The ethical guidelines embedded within these algorithms could include prioritizing human life, thus aligning the system's actions with humanitarian values. Additionally, integrating fail-safes for human intervention is crucial. This would allow human operators to override or guide the system's decisions in complex scenarios where algorithmic decision-making might not suffice. Such a blend of automated and human decision-making ensures that the system's actions remain ethically sound and adaptable to the unpredictable nature of disaster scenarios.

In the Healthcare Scenario: The ethical challenge in fog computing within healthcare revolves around maintaining patient confidentiality while processing data across a distributed network. To mitigate risks associated with data privacy and security, the implementation of advanced encryption methods is essential. These methods would ensure the security of data both in transit and at rest, safeguarding sensitive patient information against unauthorized access and breaches. Furthermore, stringent access controls are necessary to ensure that only authorized personnel have access to patient data. This not only protects patient privacy but also aligns with legal compliance requirements. In addition to these technical measures, the development of transparent policies for data usage and sharing is imperative. These policies should clearly outline how patient data is used and shared within the network, ensuring that patients' rights to privacy are respected. By doing so, these solutions uphold ethical principles such as privacy, non-maleficence (avoiding harm), and beneficence (promoting the well-being of patients).

These analyses underscore the importance of integrating ethical considerations into the design and operation of systems in swarm intelligence and fog computing. By implementing ethical decision-making algorithms, advanced encryption, stringent access controls, and transparent policies, these technologies can navigate complex ethical landscapes, ensuring that their benefits are harnessed without compromising ethical principles and societal values.

9.6.3 Lessons Learned and Best Practices

This section synthesizes key insights and actionable guidelines derived from the analysis of real-world scenarios in swarm intelligence for disaster relief and fog computing in healthcare. These lessons and practices are instrumental in guiding future developments in these fields to ethically align with societal needs and expectations.

9.6.3.1 Key Lessons Learned

Ethical Integration in Design and Operation: A primary lesson is the critical importance of integrating ethical considerations into both the design and operational phases of swarm intelligence and fog computing systems. Ethical challenges should be anticipated and addressed right from the conceptualization stage, ensuring that systems are inherently aligned with ethical norms.

Transparency is Crucial: In both scenarios, the necessity for transparency in algorithms and data policies is highlighted. Transparency not only fosters trust among users but also facilitates scrutiny and accountability, ensuring that systems operate within agreed-upon ethical boundaries.

Robust Security Measures: The importance of robust security measures, especially in scenarios involving sensitive data like healthcare, is underscored. Protecting data against unauthorized access and breaches is essential to maintaining user trust and complying with legal and ethical standards.

9.6.3.2 Best Practices

Developing Transparent Algorithms: In designing algorithms for swarm intelligence, prioritize transparency and accountability. This could involve creating interfaces or reports that explain decision-making processes and outcomes, making the system's operations understandable and justifiable to users and stakeholders.

Implementing Advanced Security Protocols: Particularly in cloud computing applications like healthcare, employing advanced encryption and stringent access controls is a best practice. Ensuring data integrity and confidentiality is non-negotiable in these sensitive applications.

Stakeholder Involvement in Ethical Decision-Making: Actively involve stakeholders, such as healthcare professionals and patients in healthcare scenarios or disaster relief experts in swarm intelligence applications, in the development process. Their insights can ensure that the systems are not only technically sound but also ethically aligned and practically useful.

Regular Ethical Audits and Reviews: Conducting regular ethical audits and reviews of the systems can help in the early identification and rectification of potential ethical issues. This proactive approach ensures continual alignment with ethical standards.

Adaptive and Responsive Systems: Design systems that are capable of adapting to new ethical challenges and societal expectations. This includes incorporating mechanisms for system evolution and updates in response to emerging ethical considerations.

In summary, the lessons learned and best practices derived from these scenarios emphasize the need for a holistic approach to developing swarm intelligence and fog computing systems. This approach should prioritize ethical considerations, transparency, security, stakeholder involvement, and adaptability to ensure that these advanced technologies are developed and deployed in ways that are beneficial, responsible, and aligned with human values.

9.7 Future Directions and Challenges

In this section, the focus shifts to the emerging trends and developments in swarm intelligence and fog computing, along with the potential future ethical challenges these fields may face. This section also emphasizes the critical role of continuous ethical assessment and adaptation in response to these evolving technologies.

9.7.1 Emerging Trends in Swarm Intelligence and Fog Computing

Emerging trends in the fields of swarm intelligence and fog computing are shaping the future of these technologies, driving them towards greater sophistication and efficiency. In swarm intelligence, a notable trend is the integration of artificial intelligence (AI) and machine learning (ML) techniques. This integration is revolutionizing how swarm systems operate, enhancing their autonomous decision-making capabilities. By incorporating AI and ML, swarm intelligence systems are becoming more adept at processing complex data, learning from past experiences, and making more informed decisions. This evolution is expected to lead to the development of swarm systems capable of performing intricate tasks with minimal human intervention, thereby increasing efficiency in various applications, such as environmental monitoring, search and rescue operations, and agricultural management.

Another significant emerging trend is seen in fog computing, particularly the move towards edge AI. This trend involves shifting AI processing closer to the point where data is generated, which is a major shift from traditional cloud computing paradigms. By processing data at the network's edge, edge AI in fog computing environments significantly reduces latency, a critical factor in applications requiring real-time processing. This reduction in latency is particularly transformative in fields such as autonomous vehicle operation and smart city infrastructure, where milliseconds can make a substantial difference in decision-making processes. Additionally, edge AI enhances data privacy and security, as less sensitive information needs to be transmitted over long distances to central servers.

Both of these trends, the integration of AI and ML in swarm intelligence and the advent of edge AI in fog computing are not only technological advancements but also harbingers of transformative changes in how data-driven decisions are made and executed. They signify a move towards more autonomous, efficient, and responsive systems capable of handling the increasing demands of modern technological applications.

9.7.2 Anticipating Future Ethical Challenges

As we look towards the future and the advancements in swarm intelligence and fog computing, it is imperative to consider and prepare for the potential ethical challenges that these technologies may pose. The increasing autonomy in swarm intelligence systems and the decentralization in fog computing bring with them a complex array of ethical considerations that need to be addressed proactively. One of the key ethical challenges with the rise of autonomous swarm intelligence systems is the issue of accountability and responsibility. As these systems become more self-sufficient in making decisions, determining who is responsible for the outcomes of these decisions becomes increasingly complicated. When these systems' decisions have significant repercussions, as in emergency response or healthcare scenarios, the complexity is even more pronounced. This raises crucial questions. Who is held accountable if an autonomous swarm system makes a decision that leads to unforeseen negative consequences? When a group of autonomous agents rather than a single entity makes decisions, how do we establish a framework for responsibility?

In the realm of fog computing, the decentralization of data processing presents significant concerns regarding data security and privacy. As data is processed and stored across numerous nodes in a network, it potentially becomes more vulnerable to security breaches and unauthorized access. Ensuring the confidentiality and integrity of data in such a distributed environment is a daunting challenge and necessitates the development of robust security protocols and systems. Another concern is the transparency and understanding of decision-making processes in both swarm intelligence and fog computing systems. As these technologies become more sophisticated, their operations may also become opaque, making it difficult for users and stakeholders to understand how decisions are made. This lack of transparency can erode public trust, which is essential for the widespread acceptance and integration of these technologies into society. Additionally, as swarm intelligence and fog computing systems become more ingrained in critical infrastructure and everyday life, there is a risk that they might inadvertently contribute to social inequalities or biases. This is particularly concerning in applications that directly impact individuals, such as predictive policing or personalized healthcare. It is vital to ensure that these systems are designed and operated in ways that are fair, unbiased, and equitable.

In summary, the future advancements in swarm intelligence and fog computing bring with them a spectrum of ethical challenges, ranging from accountability and responsibility in autonomous systems to ensuring privacy, security, transparency, and equity. Addressing these challenges requires a multidisciplinary approach involving technologists, ethicists, policymakers, and other stakeholders to develop robust frameworks and guidelines that ensure these technologies are developed and deployed responsibly.

9.7.3 The Role of Continuous Ethical Assessment and Adaptation

The rapid evolution of swarm intelligence and fog computing technologies necessitates a dynamic approach to ethical oversight, one that involves continuous assessment and adaptation. This approach recognises that as these technologies advance, the ethical landscapes in which they operate also change, bringing new challenges and considerations to the forefront. Continuous ethical assessment means regularly evaluating the impact of these technologies on society and individual rights. This is not a one-time activity but an ongoing process that aligns with the pace of technological development. It involves scrutinizing how these systems interact with users and their environments, as well as the broader societal implications of their deployment. As these technologies evolve, so too must the ethical frameworks and guidelines that govern them. This evolution is crucial to ensuring that the guidelines remain relevant and effective in addressing new ethical challenges as they arise. Central to this process is the conduct of regular audits. These audits should assess not only the technical performance of the systems but also their compliance with ethical standards. They should examine aspects such as data privacy, security, fairness, and transparency, ensuring that the systems adhere to established ethical norms and legal requirements.

Stakeholder engagement is another critical component of continuous ethical assessment. This involves bringing together a diverse group of stakeholders, including technologists, ethicists, policymakers, users, and representatives from affected communities. Engaging with a wide range of perspectives is essential for a comprehensive understanding of the ethical implications of these technologies. It helps in identifying potential issues that might not be evident from a purely technical standpoint and ensures that the solutions developed are well-rounded and considerate of various societal needs and values. Interdisciplinary research is also vital in this context. Collaborations between technologists, ethicists, social scientists, and legal experts can provide a holistic view of the challenges at hand. This research can help in developing new ethical frameworks and guidelines that are well-suited to the complexities of swarm intelligence and fog computing technologies.

In summary, the role of continuous ethical assessment and adaptation in the fields of swarm intelligence and fog computing is fundamental. Dynamic ethical frameworks that develop in tandem with these technologies are necessary to guide them. Regular audits, diverse stakeholder engagement, and interdisciplinary research are key to ensuring that these technologies are developed and deployed in a manner that is responsible, ethical, and aligned with societal values and individual rights.

9.8 Conclusion

In conclusion, this chapter has extensively explored the ethical dimensions of swarm intelligence and fog computing, two rapidly advancing technological domains. Key points highlighted throughout the chapter include the intricate balance between technological innovation and ethical considerations, the need for ethical frameworks to guide the development and application of these technologies, and the challenges and solutions associated with implementing these frameworks in real-world scenarios. The chapter began by defining swarm intelligence and fog computing, setting the stage for understanding their potential and complexities. It then delved into the technical foundations, underscoring core principles and architectures. The discussion on ethical considerations brought to light various challenges, such as privacy concerns, data security, accountability in decentralized systems, and fairness in algorithmic decision-making. In addressing these challenges, the chapter emphasized the importance of ethical frameworks and guidelines, suggesting how existing frameworks can be adapted to these specific technologies. It also highlighted the pivotal role of regulatory bodies and standards in shaping ethical practices. The implementation of ethical practices was discussed, advocating for strategies in ethical data management, transparency and accountability in algorithms, and the adoption of ethical design principles in swarm intelligence systems. Real-world case studies provided practical insights into these concepts, illustrating ethical dilemmas and resolutions in diverse scenarios. Throughout the chapter, the importance of ethical considerations in the advancement of swarm intelligence and fog computing was a recurring theme. As these technologies continue to integrate into various sectors, their ethical implications become increasingly significant. Ensuring that they align with societal values and ethical norms is not just a regulatory requirement but a fundamental aspect of responsible innovation. Finally, the chapter concludes with a call to action for researchers and practitioners. It encourages further research into the ethical aspects of these technologies, fostering a culture of continuous dialogue and discussion. Practitioners are urged to apply these ethical considerations in real-world applications, ensuring that the development and deployment of swarm intelligence and fog computing technologies are guided by a strong ethical compass. This collaborative effort between researchers, practitioners, and policymakers is crucial in steering these technological advancements towards a future that is not only innovative but also ethically sound and socially responsible.

Chapter 10

A Robust DIW Approach for UAV Image Transmission Using PSMD Encryption

Jose Santamaria Lopez,[1] *A. Sai Vasanthi,*[2] *Manisha Velaga,*[2]
Akula Narendra Kumar,[2] *B. Kavya,*[2] *Vinsha Shaik,*[2] and *Priyanka S.*[1,*]

Unmanned Aerial Vehicles (UAVs) are also known as drone technology, which grew significantly over the previous decade due to the large number of users. This technique has enormous potential in sectors such as medical care, the agricultural sector, and forestry. Nonetheless, the conveyance of digital images through drones encounters a number of problems and potential threats due to inadequate safety precautions. To overcome this issue, Digital Image Watermarking (DIW) has emerged as a feasible option for securely sending digital images in the unmanned aerial vehicle (UAV) sector. Achieving resilience, security, and imperceptibility simultaneously in a watermarking system poses a significant challenge. In response to these challenges, this study introduces a robust digital image watermarking methodology for the hybrid DWT-SVD field of a novel four-tier encryption method, PSMD (Partitioning, Substituting, Merging, Division), based

[1] Head of Research Group, Applied Computational Engineering, University of Jaen, Spain.
[2] Department of Computer Science and Engineering, SRM University, AP, India.
* Corresponding author: priyanka.s@srmap.edu.in

on symmetric cryptography to enhance UAV image security. This approach leverages Fibonacci and prime number series for the generation of two random keys, ensuring robust security with minimal computational overhead. The experimental outcome demonstrates that imperceptibility is high for lena grayscale image with 43.63db and 50.64db for color images of Image 5. Intuitive and empirical experiments demonstrate that the proposed encryption technique is highly secure.

10.1 Introduction

Unmanned aerial vehicle (UAV) usage is fast increasing due to the aerial device's outstanding image acquisition capabilities. The ability for gathering visuals is constantly improving, resulting in more diverse and huge visual information and sophisticated applications in 5G networks and IoT elements. UAVs are involved in agriculture, forestry-related emissions, archaeological studies, architecture, ecological tracking, healthcare appraisal, emergency response, and traffic intelligence. Furthermore, UAVs rely primarily on crucial parts, specifically their communications systems. The authors in [225] provide a wireless digital communication system for UAVs to assist drone operators and to achieve their aims. Information/images from hot spots areas are transmitted to base stations via multi-layer protocols in UAV connection. The application layer communicates primarily using wireless protocols like [225] IEEE 802.11 and IEEE 802.15.4.

Researchers are showing a growing interest in UAV or drone imaging systems because of its amazing spatial high-resolution capacity, low acquisition costs, minimal maintenance requirements, and real-time data sharing [238]. In UAVs' critical function, especially in the Internet of Drones (IOD), security concerns have arisen regarding both UAV connectivity (UAV-to-UAV and UAV-to-control center) and sensing data [155]. Security challenges in UAV communication include tracking, denial of service, control tampering among others [21]. Authentication, non-repudiation, and integrity verification are required for secure UAV communication, particularly when delivering digital photos. Cryptography, steganography, and digital image watermarking (DIW) all address security concerns, with DIW being a popular option [228]. DIW protects authenticity, security, and copyright by embedding watermarks in spatial or frequency domains [73]. Researchers have developed hybrid transforms to achieve imperceptibility, robustness, and capacity in DIW schemes [73]. Maintaining all of these properties in an individual DIW system is challenging and often involves a trade-off. Researchers have developed an optimum scaling factor for embedding to address trade-offs in characteristics [33]. As previously mentioned, data security is a crucial concern for UAVs. UAVs are secured using classic techniques like coding, hash functions, and asymmetric encryption. However, most encryption algorithms have substantial processing costs [98]. Subsequently, achieving resilience, security, and imperceptibility in a watermarking scheme is a difficult task. This work presents a safe digital image watermarking approach using the hybrid DWT-SVD sector to overcome aforementioned challenges.

10.2 Motivation and Contribution

To provide secure UAV connection, images must be authenticated during data transmission between base stations and UAVs. Protecting UAV mission metadata includes location name, drone pilot, and meteorological conditions. UAV security measures include system ID verification, signal spoofing, and unauthorized access. Preventing unauthorized exposure of confidential insights will interchange between UAVs and the base center, which could lead to sensitive flight mission information leaks, including Telemetry information and control instructions. This study proposes a DIW approach to authenticate UAVs and limit their participation in flight missions. The main contribution of the proposed method is as follows:

1. The PSMD approach is utilized for both encryption and decryption, providing robust watermark security at a minimal computational cost. Two symmetric random secret keys are generated at random for cipher image output.
2. The proposed Digital Image Watermarking technology is suitable for secure UAV image transference copyright enforcement, military-related, medical, and identifying applications.

10.3 Related Works

UAVs are gradually making a significant contribution to our daily lives, organizations, and governments. UAVs are commonly employed in the following domains: safety and control, military operations, air reconnaissance, forest and water search and rescue, firefighting, disaster monitoring and administration (such as floods and earthquakes), surveys, and 3D environment modelling, copyright protection and authentication of digital images are vital for many web-based applications, including IoT, IoMT, Telemedicine, Blockchain, social media, and UAV. It is critical to implement identification and security measures to protect images that are transmitted over the internet. Researchers have developed various Digital Image Watermarking (DIW) techniques that prioritize authentication process, copyright enforcement, imperceptibility, and resilience. The existing literature emphasizes DIW techniques primarily rely on spatial and transform schemes for transmitting images across digital communication channels. Su et al. [236] introduced a spatial domain technique that relies on the 2D-DFT watermarking system. This system employs a quantization technique to embed and extract geographical data using the DC coefficient. The suggested technique has a short runtime and weak watermark security. Similarly, Abrahim et al. [7] suggested a color watermarking approach for spatial domains. This approach had good visual quality, but it lacked security for the watermark. Spatial watermarking is invisible but less strong. To attain resilience, researchers mostly investigated frequency domain techniques.

Tewari et al. [242] proposed a robust DCT-based watermarking system that embeds the watermark in the mid-frequency band to improve robustness against image processing attacks. Fares et al. [76] suggested a Fast Fourier Transform technique for color picture transmission, but it does not support watermark protection. Researchers frequently experiment with hybrid transform watermarking approaches to solve imperceptibility and robustness concerns. Lai et al. [260] introduced a hybrid SVD and IWT technique that divided the image into segments, applied IWT, selected low frequencies, and used SVD decomposition. The encrypted watermark is then inserted using an evolutionary process that optimizes robustness and imperceptibility.

Alzahrani et al. [24] proposed a DWT-SVD method without considering security concerns. Yang et al. [254] proposed a DWT-SVD technique for reliable image transmission that ensures watermark security by compression and encryption with a two-dimensional chaotic map. This strategy incurs a substantial computational cost for better security. Begum [41] proposed a three-level DCT-DWT-SVD technique with initial watermark encryption utilizing an Arnold map to assure high security. Nazir et al. [172] developed an enhanced evolution fruit-fly optimization technique in the hybrid DWT-HbD-SVD sector for robust dissemination, attaining watermark security by DNA and 4D chaotic map use, albeit at a higher computational cost. Kothari et al. [243] suggested a blind hybrid NSCT-RDWT DIW system that prioritizes robustness above watermark security. Zear et al. [256] presented a secure LWT-DCT-SVD DIW system that employs MD-5 hashing. IWT-SVD with evolutionary algorithm optimization, but ignored watermark security. Many described DIW methods place little focus on watermark security, in addition to demonstrating lower degrees of security. Notably, schemes developed by [101, 172, 254, 256] provided great security at a higher cost.

10.4 Proposed Work

In this proposed scheme, a DIW technique for the secure transmission of UAV photos is developed. A cover image, denoted as CI with dimensions M x N, is paired with a binary image, referred to as WM, having dimensions A x B are encrypted and embedded in the cover image. The proposed work proceeds as follows:

(1) Watermark encryption and decryption.
(2) Watermark embedding and extraction.

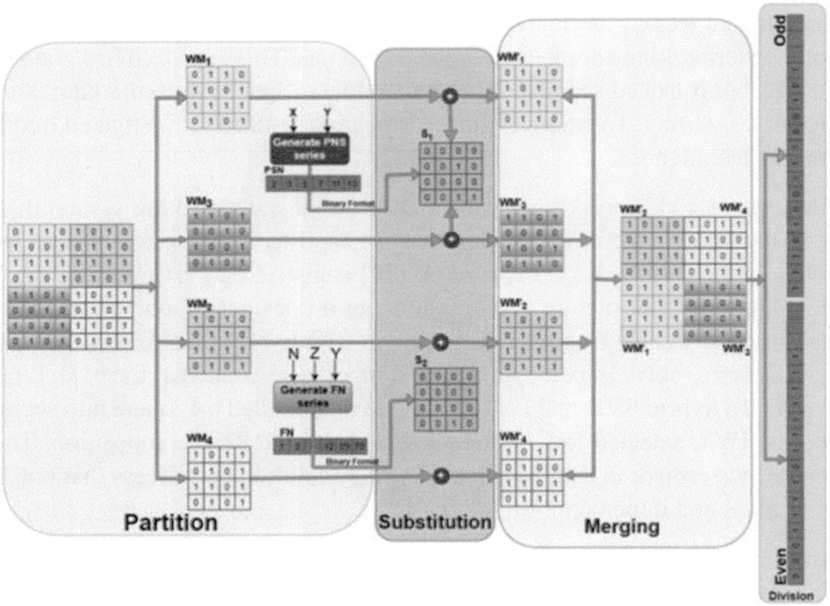

Figure 10.1: Watermark encryption process.

10.4.1 Watermark Encryption and Decryption

Information that is private can be protected with a watermark. To do that, encrypting it before embedding it into the cover image. Here, greater watermark security is ensured by the use of symmetric cryptography. For watermark encryption and decryption, the same key is used. The PSMD (Partitioning, Substitution, Merging, and Division) technique is used in four stages of the process. Randomly generated secret keys are used to get encrypted watermarks.

Watermark Encryption The encryption technique is implemented using the four PSMD layers.
1) Partitioning
2) Merging
3) Combining
4) Re-substitution

In Figure 10.1, the complete watermark encryption procedure is demonstrated using a numerical example.

Partitioning:

This is the first level of encryption. In this process, Watermark(WM) is divided into four sub-samples (WM_1, WM_2, WM_3, WM_4).

$$\sum_{p=1}^{A/2}\sum_{q=1}^{B/2} WM_1(p,q) = \sum_{r=1}^{A/2}\sum_{s=1}^{B/2} WM(s,t) \tag{10.1}$$

$$\sum_{p=1}^{A/2}\sum_{q=1}^{B/2} WM_2(p,q) = \sum_{r=1}^{A/2}\sum_{s=B/2+1}^{B} WM(s,t) \tag{10.2}$$

$$\sum_{p=1}^{A/2}\sum_{q=1}^{B/2} WM_3(p,q) = \sum_{r=A/2+1}^{A}\sum_{s=1}^{B/2} WM(s,t) \tag{10.3}$$

$$\sum_{p=1}^{A/2}\sum_{q=1}^{B/2} WM_4(p,q) = \sum_{r=A/2+1}^{A}\sum_{s=B/2+1}^{B} WM(s,t) \tag{10.4}$$

where AB is the watermark's size and WM is the watermark, and (p,q), and (r,s) are pixel positions(WM_1, WM_2, WM_3, WM_4).

Substitution:

The sub-samples pixel bits are encrypted using the XOR process by using randomly generated binary secret keys (S1, S2) which means substituting pixel bits in the relevant sub-samples. The prime number sieve series (PNS) and the generalised distance Fibonacci number series (FN), respectively, are used to create the secret keys (S1, S2) [230, 231]. The procedure for creating PNS using the seed values X and Y and recurrence relation for generalized distance Fibonacci series with seed values N, Z, and Y are described in Sorenson (1990) [230] and Falcon (2014) [74] respectively. In this case, $X \geq 1$ is the start number. N is any natural integer, $N \geq 1$, and Y is the range of the series,$1 \leq Y \leq (A2 \times B2)$. Z is the separation in series numbers, $Z \geq 1$. The watermark has dimensions $(A \times B)$, while the keyspace of the two keys is $(A4 \times B4)$, constituting one-fourth of the watermark's size. As per the proposed plan, FN produces distinct Fibonacci sequences corresponding to various values of N and Z. It strengthens the encryption scheme's security. In order to perform the decryption process, the recipient receives the secret keys X, Y, N, and Z securely via services provided by third parties. After the PNS and FN series' generation, every random number in both sequences is transformed into an 8-bit binary format, resulting in the binary values S1,S2, whose sizes match $(A2 \times B2)$. The XoR procedure is used to encrypt the subsamples (S_1, WM_2, WM_3). This is the same for subsamples S_2, WM_2,WM_4.

Merging:

Enhance the security of the watermark, the sub samples undergo encryption and are combined in a jumbled fashion, as illustrated in Figure 10.1, so that WM_1position is replaced by WM_2, WM_3position is replaced by WM_4, WM_3 position is replaced by WM_1, WM_4 position is replaced by WM_3. The sub-samples are merged using the following equations to obtain the encrypted watermark (WM').

$$\sum_{r=1}^{A/2}\sum_{s=1}^{B/2} WM'(r,s) = \sum_{p=1}^{A/2}\sum_{q=1}^{B/2} WM_2(p,q) \tag{10.5}$$

$$\sum_{r=1}^{A/2}\sum_{s=B/2+1}^{B} WM'(r,s) = \sum_{p=1}^{A/2}\sum_{q=1}^{B/2} WM_4(p,q) \tag{10.6}$$

$$\sum_{r=A/2+1}^{A}\sum_{s=1}^{B/2} WM'(r,s) = \sum_{p=1}^{A/2}\sum_{q=1}^{B/2} WM_1(p,q) \tag{10.7}$$

$$\sum_{r=A/2+1}^{A}\sum_{s=B/2+1}^{B} WM'(r,s) = \sum_{p=1}^{A/2}\sum_{q=1}^{B/2} WM_3(p,q) \tag{10.8}$$

Division:

Using the following equations, WM' is divided into two vectors: one consisting of even elements.

$$\sum_{k=1}^{(A*B)/2} WM_e'(k) = \begin{cases} \sum_{r=1}^{A}\sum_{s=1}^{B} WM'(r,s), & ifmod(s,2)=0 \\ ignored, & Otherwise \end{cases} \tag{10.9}$$

$$\sum_{k=1}^{(A*B)/2} WM_o'(k) = \begin{cases} \sum_{r=1}^{A}\sum_{s=1}^{B} WM'(r,s), & ifmod(s,2)\neq 0 \\ ignored, & Otherwise \end{cases} \tag{10.10}$$

where (r,s), k are the pixel index positions, and the dimensions of the encrypted watermark, denoted as M are $A \times B$. As a result, the suggested encryption method provides high levels of safety at low costs.

Watermark Decryption: Decryption of a watermark involves the elimination or deciphering of a watermark embedded in digital content, usually images or videos, to unveil the initial, unmarked version.

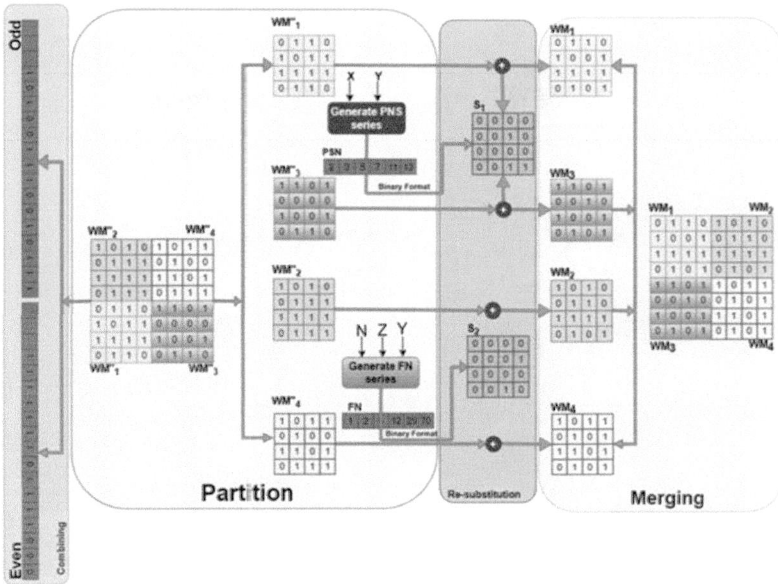

Figure 10.2: Watermark decryption process.

It typically follows mainly 4 stages:
1) Combining
2) Partitioning
3) Re-substitution
4) Merging.

Combining:

The extracted $WM''(e)$ and $WM''(o)$ are merged using the given relationship to start the decryption process.

$$\sum_{t=1}^{A}\sum_{u=1}^{B} WM''(t,u) = \begin{cases} \sum_{r=1}^{(A\times B)/2} WM_e''(r), if mod(u,2)=0 \\ \sum_{s=1}^{(A\times B)/2} WM_o''(s), Otherwise \end{cases} \tag{10.11}$$

The extracted watermark vectors are shown here as $WM''(e)$ and $WM''(o)$ With (r, s) and (t, u) standing for pixel index locations, $A \times B$ represents the size of WM''. This combination yields merged watermark WM''.

Partitioning:

Following combination, WM'' is divided into 4 sub-sampled parts using the equations from 10.12 to 10.15. The resulting sub-sampled parts are WM_1'', WM_2'',

WM_3'' and WM_4''

$$\sum_{p=1}^{A/2}\sum_{q=1}^{B/2} WM_2''(p,q) = \sum_{r=1}^{A/2}\sum_{s=1}^{B/2} WM''(r,s) \tag{10.12}$$

$$\sum_{p=1}^{A/2}\sum_{q=1}^{B/2} WM_4''(p,q) = \sum_{r=1}^{A/2}\sum_{s=B/2+1}^{B} WM''(r,s) \tag{10.13}$$

$$\sum_{p=1}^{A/2}\sum_{q=1}^{B/2} WM_3''(p,q) = \sum_{r=A/2+1}^{A}\sum_{s=B/2+1}^{B} WM''(r,s) \tag{10.14}$$

$$\sum_{p=1}^{A/2}\sum_{q=1}^{B/2} WM_1''(p,q) = \sum_{r=A/2+1}^{A}\sum_{s=1}^{B/2} WM''(r,s) \tag{10.15}$$

where C $\times$ D is the size of the WM'' obtained watermark, and (p,q), (r,s) are the pixel index areas. The watermark segments that have been retrieved are WM_1'', WM_2'', WM_3'' and WM_4''.

Re-substitution:

Using XOR operation, the randomly generated and S1, S2 secret keys are then used to decode the pixel bits of the subsampled values. Receiver and the encryption technique both make use of arbitrary seed values that are secretly communicated. These values are used for both HASL1 and HASL2 creation. After that, S1 and S2 are produced by creating and converting the associated PNS and FN random series into binary representation. Once subsampled values WM_{D1},WM_{D2},WM_{D3} and WM_{D4} are obtained, they are given to the following processing step.

Merging:

Using the below equations, the decrypted sub-sampled values are combined to obtain the final decrypted watermark image WM_D.

$$\sum_{r=1}^{A/2}\sum_{s=1}^{B/2} WM_D(r,s) = \sum_{p=1}^{A/2}\sum_{q=1}^{B/2} WM_{D1}(p,q) \tag{10.16}$$

$$\sum_{r=1}^{A/2}\sum_{s=B/2+1}^{B} WM_D(r,s) = \sum_{p=1}^{A/2}\sum_{q=1}^{B/2} WM_{D2}(p,q) \tag{10.17}$$

$$\sum_{r=A/2+1}^{A}\sum_{s=1}^{B/2} WM_D(r,s)=\sum_{p=1}^{A/2}\sum_{q=1}^{B/2} WM_{D3}(p,q) \tag{10.18}$$

$$\sum_{r=A/2+1}^{A}\sum_{s=B/2+1}^{B} WM_D(r,s)=\sum_{p=1}^{A/2}\sum_{q=1}^{B/2} WM_{D4}(p,q) \tag{10.19}$$

where $A \times B$ represents the WM's size, (p,q), (r,s) are the pixel index regions, and the decrypted watermark values are WM_{D1}, WM_{D2}, WM_{D3} and WM_{D4}. A merged watermark is WM_D.

10.4.2 Watermark Embedding and Extraction

In our proposed DIW scheme, we utilize a synergistic approach, combining SVD and DWT for watermark embedding and extraction. DWT effectively identifies subbands for precise watermark integration, capitalizing on its ability to modify specific coefficients without altering the entire image. Despite the computational efficiency of spatial domain techniques, we opt for a transform domain approach to enhance robustness, critical for UAV image authentication. The watermark is incorporated into the frequency bands (LL, LH, HL, and HH) of the DWT-decomposed cover image through adjustments to pixel intensity values within the frequency domain. After embedding, the sub-bands are merged using the inverse process, ensuring lossless recovery. Although DWT is vulnerable to certain attacks, our proposed scheme addresses this by combining DWT with SVD. SVD, known for stability and resistance to geometrical attacks, further fortifies the system against cropping, rotation, resizing, and common image manipulations [22]. Despite the computational inefficiency of hybrid DWT-SVD, this approach is recommended for its effectiveness in countering various attacks, as depicted in the watermark embedding process block diagram in Figure 10.3.

Watermark Embedding:

In the suggested method, the initial DWT transformation applied to TC leads to the formation of four subbands: LL, LH, HL, and HH. While LL contains essential image details, insertions in this subband may alter the original content. The HH subband, characterized by high frequencies, is imperceptible but less resilient.The LH and HL subbands, representing intermediate frequencies, are partitioned into 4×4 blocks α_{LH},α_{HL} for $WM'(e)$, $WM'(O)$ embedding. SVD is then applied to LH and HL, decomposing them into U_{LH}, S_{LH}, V_{LH}, U_{HL}, S_{HL},V_{HL} sub-matrices where U_{LH}, U_{LH} represents the column orthogonal matrices, S_{LH}. S_{HL} represents singular diagonal matrices and V_{LH}, V_{HL} represents the row orthogonal matrices. Although embedding in singular diagonal matrix improves visual quality, it brings about an issue of false positive errors [152]. To overcome

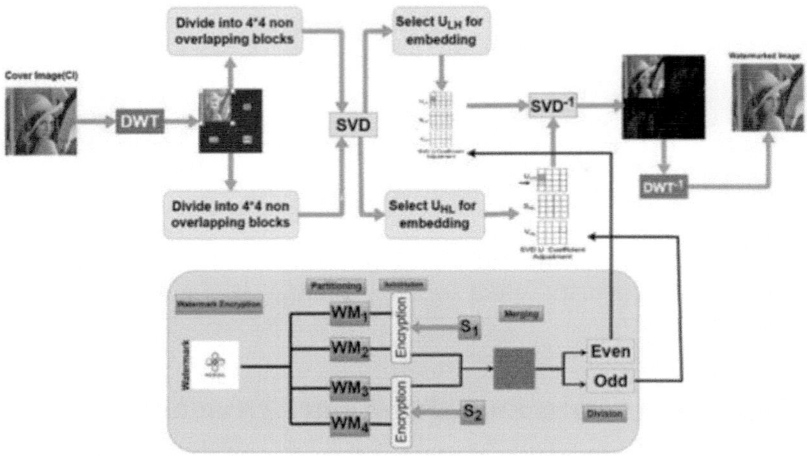

Figure 10.3: Block diagram for watermark embedding and encryption process.

this, our proposed scheme utilizes the SVD of the U matrix (U_{LH}, U_{HL}) to remove false positives and reduce side information overhead. The unique property of the SVD of U matrix allows us to embed one watermark bit in each U_{LH} and U_{HL} matrix through coefficient adjustment, minimizing image distortion [49]. The procedure involving SVD and the adjustment of U-Coefficients is explained as follows.

SVD, U-Coefficient-adjustment:

In the process of adjusting SVD U-Coefficients, the watermark is introduced by altering U coefficients based on the relation between $U_{LH}(2,1)$, $U_{HL}(3,1)$, $U_{HL}(3,1)$, $U_{LH}(2,1)$ in LH and HL subbands of U.The abs function ensures positive values for mathematical evaluation. The U matrix entries exhibit both negative and positive pixel values. Positive values are crucial during coefficient adjustment, as their absence complicates watermark retrieval. The algorithm investigates the correlation among the coefficients in the first column (U(1,1), U(2,1), U(3,1), U(4,1)), emphasizing the preservation of positive correlations to facilitate efficient watermark embedding. In the U matrix of 4×4 non-overlapping blocks, the correlation between adjacent coefficients depends on the difference in their magnitudes, which may be either positive or negative. A positive relationship is assigned when the difference is positive, and vice versa. Throughout the general embedding process, this relationship is consistently preserved. Notably, if one coefficient outweighs the other, the emphasis is on the positive relationship. Within the U component, the presence of a positive correlation indicates the concealment of a bit value of 1; conversely, a bit value of 0 is embedded when there is no positive relationship. The suggested approach strategically utilizes the absolute function (abs) to efficiently maintain the positive correlation between U(2,1) and (3,1), enhancing the optimization of watermark bit embedding and extraction.

Watermark Extraction:
The proposed blind watermarking scheme facilitates extraction without requiring the original image. The watermarked image undergoes DWT, with the extraction process focusing on the subbands $LH\prime$ and $HL\prime$. These subbands are subsequently partitioned into non-overlapping 4×4 blocks. Further, SVD is applied on α, decomposed into $ULH, S\prime LH, V\prime LH$, and $U\prime HL, V\prime HL, S\prime HL, U\prime LH, U\prime LH$, and are selected for the extraction. The equations 10.20 and 10.21 depict the formulas employed for extracting the even and odd components of the watermark.

$$\sum_{l=1}^{(AXB)/2} WM_e''(l,m) = \begin{cases} 1, if abs(U'LH(2,1) \geq U'LH(3,1)) \\ 0, Otherwise \end{cases} \tag{10.20}$$

$$\sum_{l=1}^{(AXB)/2} WM_o''(l,m) = \begin{cases} 1, if abs(U'HL(2,1) \geq U'HL(3,1)) \\ 0, Otherwise \end{cases} \tag{10.21}$$

where matrices $U'LH$,$U'HL$ are derived from the SVD decomposition, representing 4x4 non-overlapping blocks. In these blocks, (2, 1) and (3, 1) correspond to pixel intensity index positions. The size of WI is A X B, and WM_e'' and WM_o'' denote the even and odd watermark vectors.

Finally, to obtain the embedded watermark, the extracted WM_e'' and WM_o'' undergo decryption using the process explained in Section 3.1.

10.5 Experimental Result and Analysis

This segment provides experimental findings to demonstrate the efficacy of the suggested approach concerning various watermarking attributes like imperceptibility and robustness. The evaluation involves test cover UAV images, both grayscale and color, sourced from the [106, 90] USC-SIPI dataset (Hsieh, 1977), Kaggle datasets [90]. The images were resized to dimensions of 512×512 and 64×64, as illustrated in Figure 10.4.

10.5.1 Imperceptibility Test

Imperceptibility is required in all watermarking applications, not just for copyright protection applications. It suggests that the cover images and watermarked images are identical. Table 10.1 shows PSNR and SSIM values for 5 test Grayscale Images. PSNR is more than 35 dB for all grayscale images. Lena has the highest PSNR value. The ideal value is 1 for SSIM, 1 indicating a perfect

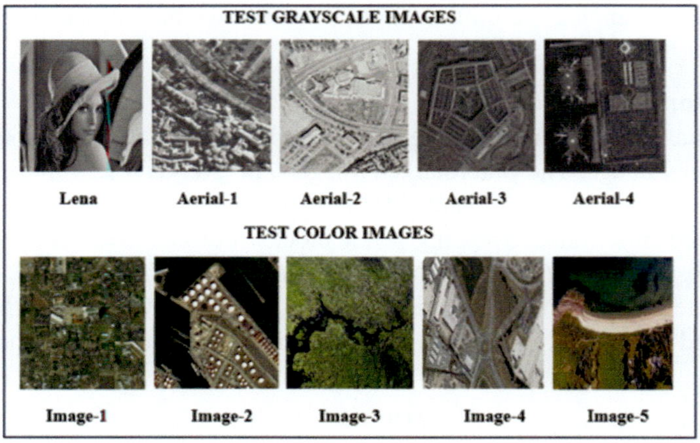

Figure 10.4: Test cover UAV images for Grayscale & Color.

Table 10.1: PSNR, SSIM for 5 Test Grayscale images.

Images	PSNR	SSIM
Lena	43.63	0.9787
Aerial 1	40.69	0.9864
Aerial 2	36.67	0.9808
Aerial 3	38.07	0.9756
Aerial 4	37.98	0.9792

match between the reconstructed image and the original image. SSIM rates of 0.97, 0.98 are reasonable for high-quality reconstruction techniques.

Table 10.2 shows PSNR and SSIM values for Color Images. Image5 has the highest PSNR value. The ideal value is 1 for SSIM, 1 indicating a perfect match between the reconstructed image and the original image. SSIM rates 0.99 for all color images.

Table 10.2: PSNR, SSIM for 5 Test Color images.

Image	PSNR	SSIM
Image1	39.78	0.9960
Image2	39.78	0.9960
Image3	44.18	0.9975
Image4	38.80	0.9936
Image5	50.64	0.9990

Table 10.3: NC and BER values for five Test Grayscale images.

Images	NC	BER
Lena	0.5575	0.4717
Aerial1	0.5953	0.4370
Aerial2	0.5920	0.4407
Aerial3	0.5778	0.4575
Aerial4	0.5724	0.4521

Table 10.4: NC and BER values for test Color images.

Images	NC	BER
Image1	0.5596	0.4697
Image2	0.5744	0.4631
Image3	0.5789	0.4631
Image4	0.5876	0.4387
Image5	0.5845	0.4524

10.5.2 Robustness Test

The USC-SIPI dataset's grayscale images, Lena, Aerial1, Aerial2, Aerial3, and Aerial4, were evaluated for Balanced Error Rate (BER) and Normalised Correlation (NC) in Table 10.3. Among the provided images, Aerial1 is the image that resembles the reference image the best among the ones that are provided in terms of robustness, having the lowest BER value of 0.4370. The next lowest BER, 0.4407, is identified in Aerial2, which is closely behind and adds to its binary error resilience. Aerial3 and Aerial4 exhibit a marginally elevated amount of error in comparison to Aerial1 and Aerial2, with respective BER values of 0.4575 and 0.4521. As the least reliable of the images taken into consideration, Lena finally displays a relatively higher BER of 0.4717. Therefore, in terms of robustness based on BER values, Aerial1 is the most robust image, followed by Aerial2, Aerial4, Aerial3, and Lena.

Among the provided images, Image 4 has the lowest BER (0.4387) as shown in Table 10.4, which indicates robustness. Next in line is Image 2, which shows a BER of 0.4631. Image 3 also shows the same BER. Image 5 exhibits the next degree of robustness, with a slightly greater BER of 0.4524. In conclusion, Image 1 has a considerably greater BER of 0.4697, suggesting a lower degree of robustness in comparison to the other images. These results show how different the colour images are in terms of quality and how susceptible they are to distortions.

The attacks with the Gaussian filter (3*3) and salt & pepper (0.0002) give the lowest NC (0.3619) and the highest NC (0.5010), respectively, showing major difference and improved similarity despite distortion. While cropping displays the lowest BER (0.4941), indicating less errors, sharpening attack produces the

Table 10.5: NC, BER under attacks for Lena image.

Attacks	NC	BER
Sharpening	0.4041	0.6271
Gaussian Filter (3*3)	0.5010	0.5291
Salt & pepper (0.0002)	0.3619	0.4271
Gaussian noise	0.4728	0.5284
Compression (60%)	0.5001	0.4962
Row cut (10)	0.5021	0.5271
Columns cut (10)	0.5041	0.5025
Cropping	0.3873	0.4941

Table 10.6: The proposed approach is compared to the other mentioned schemes.

Parameters	Zear and Singh (2022)	Zhu et al. (2022a)	Proposed Scheme
Scheme	LWT-DCT-SVD	IWT-SVD	DWT-SVD
Size of Cover image	512*512	512*512	256*256
Size of Watermark	64*64	32*32	64*64
Security	Yes	No	Yes
Computational cost	O(N)	$O(N^3)$	$O(N^3)$

highest BER (0.6271). These results demonstrate the various impacts of attacks on image resilience and similarity as shown in Table 10.5.

10.5.3 Comparative Study

This section entails a comparative examination of the suggested method against alternative techniques. The performance assessment involves a comparison with the Secure and robust color image dual watermarking based on LWT-DCT-SVD proposed by Aditi Zear and Pradeep Kumar Singh (Zear and Singh) and An optimized image watermarking algorithm based on SVD and IWT. The Journal of Supercomputing 78(1): 222–237 (Zhu et al).

Table 10.6 depicts the computations cost for the comparative schemes and the proposed scheme. The computational time for the proposed scheme is similar to schemes proposed in Zhu et al., and higher than the scheme proposed in Zear and Singh (2022) but has higher security. For comparative study, the popular test image "Lena" is considered. Graph 1 depicts that in comparison using PSNR with other techniques. The proposed technique gives a better imperceptibility. Embedding and Extraction time (in seconds) of different images is shown in Table 10.7.

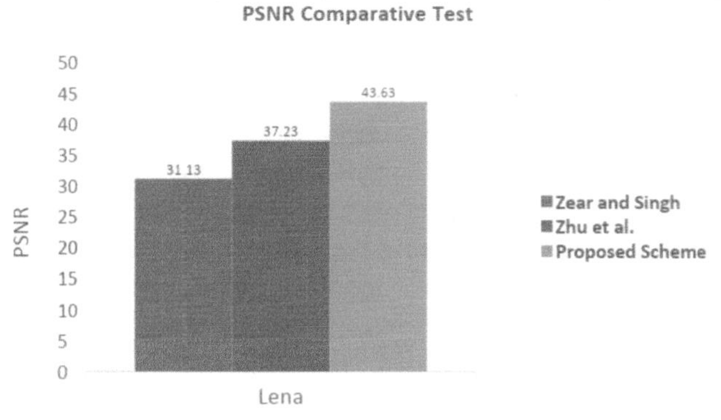

Figure 10.5: Graph1: Comparison of Peak Signal-to-Noise Ratio (PSNR) for the proposed scheme in relation to other mentioned methods using the Lena image.

Table 10.7: Embedding and Extraction time (in seconds) of different images.

Image	Embedding Time	Extraction Time
Lena	0.236803	0.090271
Aerial	0.275185	0.099864
Image1	0.397162	0.190732
Image2	0.317393	0.110072
Image3	0.387216	0.182742
Image4	0.309271	0.102864

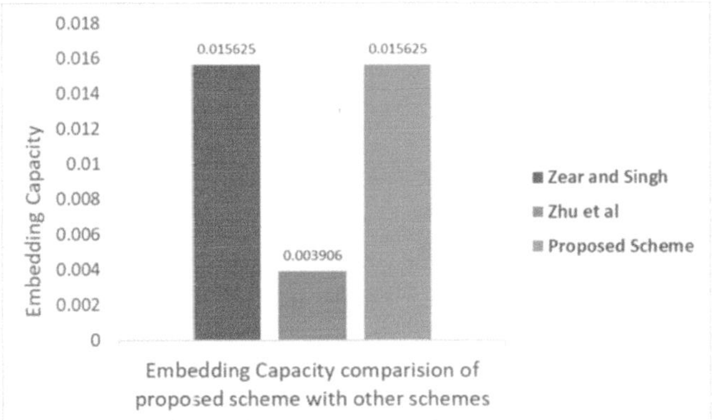

Figure 10.6: Graph2: Embedding Capacity comparison of proposed scheme with other mentioned methods.

10.6 Conclusion

This research introduces a blind and secure DIW scheme that combines optimized DWT and SVD. This scheme is specifically designed for watermarking digital images transmitted via drone applications. The proposed scheme focuses on ensuring robust transmission, security through the PSMD symmetric encryption approach, and effective watermark embedding. The effectiveness is demonstrated by comparing watermarking characteristics such as PSNR, SSIM, NC, and BER with other optimization algorithm random showcasing higher imperceptibility and robustness. Moreover, in contrast to existing DIW schemes, the proposed approach demonstrates superior imperceptibility, robustness, and security. It is applicable in practical scenarios such as UAV image transmission, safeguarding copyrights, military applications, and various other contexts. In the coming times, our objective is to develop a Digital Image Watermarking (DIW) system utilizing machine/deep learning techniques to achieve strong resistance against rotation, scaling, cropping attacks, and the application of grayscale images as watermarks. Furthermore, we aim to create an extra scheme specifically focused on preserving the integrity of Unmanned Aerial Vehicle (UAV) images.

References

[1] K Jyothsna Devi, Priyanka Singh, Muhammad Bilal, and Anand Nayyar. Enabling secure image transmission in unmanned aerial vehicle using digital image watermarking with h-grey optimization. Expert Systems with Applications, 236:121190, 2024.

[2] K Jyothsna Devi, Priyanka Singh, Hiren Kumar Thakkar, and Neeraj Kumar. Robust and secured watermarking using ja-fi optimization for digital image transmission in social media. Applied Soft Computing, 131:109781, 2022.

[3] Priyanka Singh, K Jyothsna Devi, Hiren Kumar Thakkar, Muhammad Bilal, Anand Nay- yar, and Daehan Kwak. Robust and secure medical image watermarking for edge-enabled e-healthcare. IEEE Access, 11:135831–135845, 202.

[4] `PBCLibrary,url={https://crypto.stanford.edu/pbc/}, journal={PBCLibrary}`. [Accessed 24-01-2024].

[5] Mahdi Abbasi, Mina Yaghoobikia, Milad Rafiee, Alireza Jolfaei, and Mohammad R Khosravi. Efficient resource management and workload allocation in fog–cloud computing paradigm in iot using learning classifier systems. *Computer communications*, 153:217–228, 2020.

[6] Mohamed Abdel-Basset, Hossam Hawash, Nour Moustafa, Imran Razzak, and Mohamed Abd Elfattah. Privacy-preserved learning from non-iid data in fog-assisted iot: A federated learning approach. *Digital Communications and Networks*, 2022.

[7] Jobin Abraham and Varghese Paul. An imperceptible spatial domain color image watermarking scheme. *Journal of King Saud University-Computer and Information Sciences*, 31(1):125–133, 2019.

[8] Stuart M Adams and Carol J Friedland. A survey of unmanned aerial vehicle (uav) usage for imagery collection in disaster research and management. In *9th international workshop on remote sensing for disaster response*, volume 8, pages 1–8, 2011.

[9] Sharmistha Adhikari, Sangram Ray, Mohammad S Obaidat, and GP Biswas. Efficient and secure content dissemination architecture for content centric network using ecc-based public key infrastructure. *Computer Communications*, 157:187–203, 2020.

[10] Swati Agarwal, Shashank Yadav, and Arun Kumar Yadav. An architecture for elastic resource allocation in fog computing. *Int. J. Comput. Sci. Commun*, 6(2):201–207, 2015.

[11] Mamia Agbese, Rahul Mohanani, Arif Khan, and Pekka Abrahamsson. Implementing ai ethics: Making sense of the ethical requirements. pages 62–71, 2023.

[12] Awais Ahmad, Murad Khan, Anand Paul, Sadia Din, M Mazhar Rathore, Gwanggil Jeon, and Gyu Sang Choi. Toward modeling and optimization of features selection in big data based social internet of things. *Future Generation Computer Systems*, 82:715–726, 2018.

[13] Lukman Adewale Ajao and Simon Tooswem Apeh. Secure edge computing vulnerabilities in smart cities sustainability using petri net and genetic algorithm-based reinforcement learning. *Intelligent Systems with Applications*, 18:200216, 2023.

[14] Ala Al-Fuqaha, Mohsen Guizani, Mehdi Mohammadi, Mohammed Aledhari, and Moussa Ayyash. Internet of things: A survey on enabling technologies, protocols, and applications. *IEEE communications surveys & tutorials*, 17(4):2347–2376, 2015.

[15] Ala Al-Fuqaha, Mohsen Guizani, Mehdi Mohammadi, Mohammed Aledhari, and Moussa Ayyash. Internet of things: A survey on enabling technologies, protocols, and applications. *IEEE communications surveys and tutorials*, 17(4):2347–2376, 2015.

[16] Irfan Alam et al. A novel authentication scheme for group based communication for iot oriented infrastructure in smart cities. 2022.

[17] Irfan Alam and Manoj Kumar. A novel protocol for efficient authentication in cloud-based iot devices. *Multimedia Tools and Applications*, 81(10):13823–13843, 2022.

[18] Irfan Alam and Manoj Kumar. A novel authentication protocol to ensure confidentiality among the internet of medical things in covid-19 and future pandemic scenario. *Internet of Things*, 22:100797, 2023.

[19] Irfan Alam and Manoj Kumar. Various elements of analysis of authentication schemes for iot devices: A brief overview. *Intelligent Data Analytics, IoT, and Blockchain*, pages 219–228, 2024.

[20] Wedad Alawad, Nadhir Ben Halima, and Layla Aziz. An unmanned aerial vehicle (uav) system for disaster and crisis management in smart cities. *Electronics*, 12(4):1051, 2023.

[21] Moatsum Alawida, Je Sen Teh, and Wafa'Hamdan Alshoura. A new image encryption algorithm based on dna state machine for uav data encryption. *Drones*, 7(1):38, 2023.

[22] Musrrat Ali, Chang Wook Ahn, and Millie Pant. A robust image watermarking technique using svd and differential evolution in dct domain. *Optik*, 125(1):428–434, 2014.

[23] Norah Alsaeed, Farrukh Nadeem, and Faisal Albalwy. A scalable and lightweight group authentication framework for internet of medical things using integrated blockchain and fog computing. *Future Generation Computer Systems*, 151:162–181, 2024.

[24] Ali Alzahrani et al. Enhanced invisibility and robustness of digital image watermarking based on dwt-svd. *Applied Bionics and Biomechanics*, 2022, 2022.

[25] Yehia I Alzoubi, Valmira H Osmanaj, Ashraf Jaradat, and Ahmad Al-Ahmad. Fog computing security and privacy for the internet of thing applications: State-of-the-art. *Security and Privacy*, 4:e145, 2021.

[26] Safwa Ameer, James Benson, and Ravi Sandhu. An attribute-based approach toward a secured smart-home iot access control and a comparison with a role-based approach. *Information*, 13(2):60, 2022.

[27] Adam J Andreotta, Nin Kirkham, and Marco Rizzi. Ai, big data, and the future of consent. *Ai and Society*, 37:1715–1728, 2022.

[28] Nancy A Angel, Dakshanamoorthy Ravindran, P M Durai Raj Vincent, Kathiravan Srinivasan, and Yuh-Chung Hu. Recent advances in evolving computing paradigms: Cloud, edge, and fog technologies. *Sensors*, 22:196, 2021.

[29] Naveed Anwar, Muhammad Amir Izhar, and Fawad Ahmed Najam. Construction monitoring and reporting using drones and unmanned aerial vehicles (uavs). In *The Tenth International Conference on Construction in the 21st Century (CITC-10)*, volume 8, pages 2–4, 2018.

[30] Hemant Kumar Apat, Bibhudatta Sahoo, and Prasenjit Maiti. Service placement in fog computing environment. In *2018 International Conference on Information Technology (ICIT)*, pages 272–277. IEEE, 2018.

[31] Muhammad Yeasir Arafat and Sangman Moh. A survey on cluster-based routing protocols for unmanned aerial vehicle networks. *IEEE Access*, 7:498–516, 2018.

[32] Aymen Arfaoui. Unmanned aerial vehicle: Review of onboard sensors, application fields, open problems and research issues. *Int. J. Image Process*, 11(1):12–24, 2017.

[33] Dhani Ariatmanto and Ferda Ernawan. Adaptive scaling factors based on the impact of selected dct coefficients for image watermarking. *Journal of King Saud University-Computer and Information Sciences*, 34(3):605–614, 2022.

[34] Sankalap Arora and Satvir Singh. Butterfly algorithm with lèvy flights for global optimization. pages 220–224, 09 2015.

[35] Mohammad S Aslanpour, Sukhpal Singh Gill, and Adel N Toosi. Performance evaluation metrics for cloud, fog and edge computing: A review, taxonomy, benchmarks and standards for future research. *Internet of Things*, 12:100273, 2020.

[36] Luigi Atzori, Antonio Iera, and Giacomo Morabito. The internet of things: A survey. *Computer networks*, 54(15):2787–2805, 2010.

[37] Joao Bachiega Jr, Breno Costa, Leonardo R Carvalho, Michel JF Rosa, and Aleteia Araujo. Computational resource allocation in fog computing: A comprehensive survey. *ACM Computing Surveys*, 55(14s):1–31, 2023.

[38] Fenye Bao and Ing-Ray Chen. Dynamic trust management for internet of things applications. In *Proceedings of the 2012 international workshop on Self-aware internet of things*, pages 1–6, 2012.

[39] Fenye Bao, Ray Chen, and Jia Guo. Scalable, adaptive and survivable trust management for community of interest based internet of things systems. In *2013 IEEE eleventh international symposium on autonomous decentralized systems (ISADS)*, pages 1–7. IEEE, 2013.

[40] Paulo SLM Barreto, Benoît Libert, Noel McCullagh, and Jean-Jacques Quisquater. Efficient and provably-secure identity-based signatures and signcryption from bilinear maps. In *Advances in Cryptology-ASIACRYPT 2005: 11th International Conference on the Theory and Application of Cryptology and Information Security, Chennai, India, December 4-8, 2005. Proceedings 11*, pages 515–532. Springer, 2005.

[41] Mahbuba Begum, Jannatul Ferdush, and Mohammad Shorif Uddin. A hybrid robust watermarking system based on discrete cosine transform, discrete wavelet transform, and singular value decomposition. *Journal of King Saud University-Computer and Information Sciences*, 34(8):5856–5867, 2022.

[42] Julian Bellendorf and Zoltán Ádám Mann. Classification of optimization problems in fog computing. *Future Generation Computer Systems*, 107:158–176, 2020.

[43] Salim Bitam, Sherali Zeadally, and Abdelhamid Mellouk. Fog computing job scheduling optimization based on bees swarm. *Enterprise Information Systems*, 12:373–397, 2018.

[44] Luiz Bittencourt, Roger Immich, Rizos Sakellariou, Nelson Fonseca, Edmundo Madeira, Marilia Curado, Leandro Villas, Luiz DaSilva, Craig Lee, and Omer Rana. The internet of things, fog and cloud continuum: Integration and challenges. *Internet of Things*, 3:134–155, 2018.

[45] Sriramulu Bojjagani, PV Venkateswara Rao, Dinesh Reddy Vemula, B Ramachandra Reddy, and T Jaya Lakshmi. A secure iot-based micropayment protocol for wearable devices. *Peer-to-Peer Networking and Applications*, 15(2):1163–1188, 2022.

[46] Eleonora Borgia. The internet of things vision: Key features, applications and open issues. *Computer Communications*, 54:1–31, 2014.

[47] Rajkumar Buyya and Satish Narayana Srirama. *Fog and edge computing: principles and paradigms*. John Wiley and Sons, 2019.

[48] Ali Çalhan and Murtaza Cicioğlu. Drone-assisted smart data gathering for pandemic situations. *Computers & Electrical Engineering*, 98:107769, 2022.

[49] Chin-Chen Chang, Piyu Tsai, and Chia-Chen Lin. Svd-based digital image watermarking scheme. *Pattern Recognition Letters*, 26(10):1577–1586, 2005.

[50] Dong Chen, Guiran Chang, Dawei Sun, Jiajia Li, Jie Jia, and Xingwei Wang. Trm-iot: A trust management model based on fuzzy reputation for internet of things. *Computer Science and Information Systems*, 8(4):1207–1228, 2011.

[51] Liang Chen and Jason Crampton. Risk-aware role-based access control. In *Security and Trust Management: 7th International Workshop, STM 2011, Copenhagen, Denmark, June 27-28, 2011, Revised Selected Papers 7*, pages 140–156. Springer, 2012.

[52] Younghoon Choi, Youngjun Choi, Simon Briceno, and Dimitri N Mavris. Multi-uas path-planning for a large-scale disjoint disaster management. In *2019 International Conference on Unmanned Aircraft Systems (ICUAS)*, pages 799–807. IEEE, 2019.

[53] Shahrin Chowdhury and Anca Delia Jurcut. Design requirements for secured and cost-efficient blockchain-based data exchange frameworks in healthcare. pages 474–486. IEEE, 2023.

[54] Sudipta Chowdhury, Adindu Emelogu, Mohammad Marufuzzaman, Sarah G Nurre, and Linkan Bian. Drones for disaster response and relief operations: A continuous approximation model. *International Journal of Production Economics*, 188:167–184, 2017.

[55] Erik Cuevas, Miguel Cienfuegos, Daniel Zaldívar, and Marco Pérez-Cisneros. A swarm optimization algorithm inspired in the behavior of the social-spider. *Expert Systems with Applications*, 40(16):6374–6384, 2013.

[56] Sharifah Mastura Syed Mohd Daud, Mohd Yusmiaidil Putera Mohd Yusof, Chong Chin Heo, Lay See Khoo, Mansharan Kaur Chainchel Singh, Mohd Shah Mahmood, and Hapizah Nawawi. Applications of drone in disaster management: A scoping review. *Science & Justice*, 62(1):30–42, 2022.

[57] Karl de Fine Licht and Jenny de Fine Licht. Artificial intelligence, transparency, and public decision-making: Why explanations are key when trying to produce perceived legitimacy. *AI and society*, 35:917–926, 2020.

[58] Chandramohan Dhasarathan, Mohammad Kamrul Hasan, Shayla Islam, Salwani Abdullah, Shailesh Khapre, Dalbir Singh, Abdulaziz A Alsulami, and Ali Alqahtani. User privacy prevention model using supervised federated learning-based block chain approach for internet of medical things. *CAAI Transactions on Intelligence Technology*, 2023.

[59] Chandramohan Dhasarathan, Mohammad Kamrul Hasan, Shayla Islam, Salwani Abdullah, Umi Asma Mokhtar, Abdul Rehman Javed, and Sam Goundar. Covid-19 health data analysis and personal data preserving: A homomorphic privacy enforcement approach. *Computer communications*, 199:87–97, 2023.

[60] Chandramohan Dhasarathan, M Shanmugam, Manish Kumar, Diwakar Tripathi, Shailesh Khapre, and Achyut Shankar. A nomadic multi-agent based privacy metrics for e-health care: a deep learning approach. *Multimedia Tools and Applications*, pages 1–24, 2023.

[61] Whitfield Diffie and Martin E Hellman. New directions in cryptography. In *Democratizing Cryptography: The Work of Whitfield Diffie and Martin Hellman*, pages 365–390. 2022.

[62] Ahmad Din, Meh Jabeen, Kashif Zia, Abbas Khalid, and Dinesh Kumar Saini. Behavior-based swarm robotic search and rescue using fuzzy controller. *Computers and Electrical Engineering*, 70:53–65, 2018.

[63] Weiping Ding, Mohamed Abdel-Basset, Hossam Hawash, Sara Abdel-Razek, and Chuansheng Liu. Fed-esd: Federated learning for efficient epileptic seizure detection in the fog-assisted internet of medical things. *Information Sciences*, 630:403–419, 2023.

[64] Jasenka Dizdarević, Francisco Carpio, Admela Jukan, and Xavi Masip-Bruin. A survey of communication protocols for internet of things and related challenges of fog and cloud computing integration. *ACM Computing Surveys (CSUR)*, 51(6):1–29, 2019.

[65] Marco Dorigo and Gianni Di Caro. Ant colony optimization: A new meta-heuristic. volume 2, page 1477 Vol. 2, 02 1999.

[66] Elie El Haber, Tri Minh Nguyen, Dariush Ebrahimi, and Chadi Assi. Computational cost and energy efficient task offloading in hierarchical edge-clouds. In *2018 IEEE 29th Annual International Symposium on Personal, Indoor and Mobile Radio Communications (PIMRC)*, pages 1–6. IEEE, 2018.

[67] Taher ElGamal. A public key cryptosystem and a signature scheme based on discrete logarithms. *IEEE transactions on information theory*, 31(4):469–472, 1985.

[68] Ahmed Elhadad, Fulayjan Alanazi, Ahmed I Taloba, and Amr Abozeid. Fog computing service in the healthcare monitoring system for managing the real-time notification. *Journal of Healthcare Engineering*, 2022, 2022.

[69] Milan Erdelj and Enrico Natalizio. Uav-assisted disaster management: Applications and open issues. In *2016 international conference on computing, networking and communications (ICNC)*, pages 1–5. IEEE, 2016.

[70] Milan Erdelj and Enrico Natalizio. Uav-assisted disaster management: Applications and open issues. In *2016 international conference on computing, networking and communications (ICNC)*, pages 1–5. IEEE, 2016.

[71] Milan Erdelj, Enrico Natalizio, Kaushik R Chowdhury, and Ian F Akyildiz. Help from the sky: Leveraging uavs for disaster management. *IEEE Pervasive Computing*, 16(1):24–32, 2017.

[72] Samuel Kofi Erskine and Khaled M Elleithy. Secure intelligent vehicular network using fog computing. *Electronics*, 8:455, 2019.

[73] Oleg Evsutin and Kristina Dzhanashia. Watermarking schemes for digital images: Robustness overview. *Signal Processing: Image Communication*, 100:116523, 2022.

[74] Sergio Falcon. Generalized (k, r)–fibonacci numbers. *Gen. Math. Notes*, 25(2):148–158, 2014.

[75] Wenhao Fan, Yi Su, Jie Liu, Shenmeng Li, Wei Huang, Fan Wu, and Yuan'an Liu. Joint task offloading and resource allocation for vehicular edge computing based on v2i and v2v modes. *IEEE Transactions on Intelligent Transportation Systems*, 24(4):4277–4292, 2023.

[76] Kahlessenane Fares, Khaldi Amine, and Euschi Salah. A robust blind color image watermarking based on fourier transform domain. *Optik*, 208:164562, 2020.

[77] X. Feng, J.-W Zhang, and H.-Q Yu. Mosquito host-seeking algorithm for tsp problem. *Jisuanji Xuebao/Chinese Journal of Computers*, 37:1794–1808, 08 2014.

[78] Stefka Fidanova and Stefka Fidanova. Ant colony optimization. *Ant Colony Optimization and Applications*, pages 3–8, 2021.

[79] Samuel Akwasi Frimpong, Mu Han, Edward Kwadwo Boahen, Rexford Nii Ayitey Sosu, Isaac Hanson, Otu Larbi-Siaw, and Isaac Baffour Senkyire. Recguard: An efficient privacy preservation blockchain-based system for online social network users. *Blockchain: Research and Applications*, 4(1):100111, 2023.

[80] Ahmed G Gad. Particle swarm optimization algorithm and its applications: a systematic review. *Archives of computational methods in engineering*, 29(5):2531–2561, 2022.

[81] Subramanium Ganesan, Manish Shakya, Aqueel F Aqueel, and Lakshmi M Nambiar. Small disaster relief robots with swarm intelligence routing. pages 123–127. Association for Computing Machinery, 2011.

[82] Siddarth Ganesh, Vaishnavi Gopalasamy, and NB Sai Shibu. Architecture for drone assisted emergency ad-hoc network for disaster rescue operations. In *2021 International Conference on COMmunication Systems & NETworkS (COMSNETS)*, pages 44–49. IEEE, 2021.

[83] Sebastian Garcia, Agustin Parmisano, and Maria Jose Erquiaga. Iot-23: A labeled dataset with malicious and benign iot network traffic (version 1.0. 0)[data set]. zenodo, 2020.

[84] Na Geng, Zhiting Chen, Quang A Nguyen, and Dunwei Gong. Particle swarm optimization algorithm for the optimization of rescue task allocation with uncertain time constraints. *Complex and Intelligent Systems*, 7:873–890, 2021.

[85] Ilche Georgievski and Marco Aiello. The potentials of ai planning on the edge. In *2023 IEEE International Conference on Edge Computing and Communications (EDGE)*, pages 330–336. IEEE, 2023.

[86] Mostafa Ghobaei-Arani, Alireza Souri, and Ali A Rahmanian. Resource management approaches in fog computing: a comprehensive review. *Journal of Grid Computing*, 18(1):1–42, 2020.

[87] Ananda Mohon Ghosh and Katarina Grolinger. Edge-cloud computing for internet of things data analytics: Embedding intelligence in the edge with deep learning. *IEEE Transactions on Industrial Informatics*, 17:2191–2200, 2020.

[88] Sukhpal Singh Gill, Minxian Xu, Carlo Ottaviani, Panos Patros, Rami Bahsoon, Arash Shaghaghi, Muhammed Golec, Vlado Stankovski, Huaming Wu, and Ajith Abraham. Ai for next generation computing: Emerging trends and future directions. *Internet of Things*, 19:100514, 2022.

[89] L M Gladence, V M Anu, A Anderson, I Stanley, J A J Fernando, and S Revathy. Swarm intelligence in disaster recovery. pages 1–8, 2021.

[90] A Goldbloom and J Howard. Cloud based public data set, 2010.

[91] Mohammad Goudarzi, Huaming Wu, Marimuthu Palaniswami, and Rajkumar Buyya. An application placement technique for concurrent iot applications in edge and fog computing environments. *IEEE Transactions on Mobile Computing*, 20(4):1298–1311, 2020.

[92] Jayavardhana Gubbi, Rajkumar Buyya, Slaven Marusic, and Marimuthu Swami Palaniswami. Internet of things (iot): A vision, architectural elements, and future directions. *ArXiv*, abs/1207.0203, 2012.

[93] Carlos Guerrero, Isaac Lera, and Carlos Juiz. Genetic-based optimization in fog computing: Current trends and research opportunities. *Swarm and Evolutionary Computation*, 72:101094, 2022.

[94] Chen Guo, Heng Tang, Ben Niu, and Chang Boon Patrick Lee. A survey of bacterial foraging optimization. *Neurocomputing*, 452:728–746, 2021.

[95] Harshit Gupta, Amir Vahid Dastjerdi, Soumya K Ghosh, and Rajkumar Buyya. ifogsim: A toolkit for modeling and simulation of resource man-

agement techniques in the internet of things, edge and fog computing environments. *Software: Practice and Experience*, 47(9):1275–1296, 2017.

[96] Sergio Gusmeroli, Salvatore Piccione, and Domenico Rotondi. A capability-based security approach to manage access control in the internet of things. *Mathematical and Computer Modelling*, 58(5-6):1189–1205, 2013

[97] Yasmine Harbi, Zibouda Aliouat, Allaoua Refoufi, Saad Harous, and Abdelhak Bentaleb. Enhanced authentication and key management scheme for securing data transmission in the internet of things. *Ad Hoc Networks*, 94:101948, 2019.

[98] Mohammad Kamrul Hasan, Shayla Islam, Rossilawati Sulaiman, Sheroz Khan, Aisha-Hassan Abdalla Hashim, Shabana Habib, Mohammad Islam, Saleh Alyahya, Musse Mohamed Ahmed, Samar Kamil, et al. Lightweight encryption technique to enhance medical image security on internet of medical things applications. *IEEE Access*, 9:47731–47742, 2021.

[99] Jonas D Hasbach and Maren Bennewitz. The design of self-organizing human–swarm intelligence. *Adaptive Behavior*, 30:361–386, 2022.

[100] Hiwa Omer Hassan, Sadoon Azizi, and Mohammad Shojafar. Priority, network and energy-aware placement of iot-based application services in fog-cloud environments. *IET communications*, 14(13):2117–2129, 2020.

[101] Einolah Hatami, Hamidreza Rashidy Kanan, Kamran Layeghi, and Ali Harounabadi. An optimized robust and invisible digital image watermarking scheme in contourlet domain for protecting rightful ownership. *Multimedia Tools and Applications*, 82(2):2021–2051, 2023.

[102] Timothy C. Havens, Christopher J. Spain, Nathan G. Salmon, and James M. Keller. Roach infestation optimization. In *2008 IEEE Swarm Intelligence Symposium*, pages 1–7, 2008.

[103] Samira Hayat, Evşen Yanmaz, and Raheeb Muzaffar. Survey on unmanned aerial vehicle networks for civil applications: A communications viewpoint. *IEEE Communications Surveys & Tutorials*, 18(4):2624–2661, 2016.

[104] José L. Hernández-Ramos, Antonio J. Jara, Leandro Marín, and Antonio F. Skarmeta Gómez. Dcapbac: embedding authorization logic into smart things through ecc optimizations. *International Journal of Computer Mathematics*, 93(2):345–366, 2016.

[105] Cheol-Ho Hong and Blesson Varghese. Resource management in fog/edge computing: a survey on architectures, infrastructure, and algorithms. *ACM Computing Surveys (CSUR)*, 52(5):1–37, 2019.

[106] M Hsieh. Signal and image processing institute, 2017.

[107] Jinqiang Hu, Husheng Wu, Bin Zhong, and Renbin Xiao. Swarm intelligence-based optimisation algorithms: an overview and future research issues. *International Journal of Automation and Control*, 14:656–693, 2020.

[108] Md Muzakkir Hussain, Mohammad Saad Alam, and MM Sufyan Beg. Vehicular fog computing-planning and design. *Procedia Computer Science*, 167:2570–2580, 2020.

[109] Md Muzakkir Hussain, Ahmad Taher Azar, Rafeeq Ahmed, Syed Umar Amin, Basit Qureshi, V Dinesh Reddy, Irfan Alam, and Zafar Iqbal Khan. Song: a multi-objective evolutionary algorithm for delay and energy aware facility location in vehicular fog networks. *Sensors*, 23(2):667, 2023.

[110] Md Muzakkir Hussain and MM Sufyan Beg. Code-v: Multi-hop computation offloading in vehicular fog computing. *Future Generation Computer Systems*, 116:86–102, 2021.

[111] Muzakkir Hussain, Mohammad Saad Alam, MM Sufyan Beg, and Nadeem Akhtar. Towards minimizing delay and energy consumption in vehicular fog computing (vfc). *Journal of Intelligent & Fuzzy Systems*, 38(5):6549–6560, 2020.

[112] Sundas Iftikhar, Sukhpal Singh Gill, Chenghao Song, Minxian Xu, Mohammad Sadegh Aslanpour, Adel N Toosi, Junhui Du, Huaming Wu, Shreya Ghosh, and Deepraj Chowdhury. Ai-based fog and edge computing: A systematic review, taxonomy and future directions. *Internet of Things*, 21:100674, 2023.

[113] Michael J Jacobson, Neal Koblitz, Joseph H Silverman, Andreas Stein, and Edlyn Teske. Analysis of the xedni calculus attack. *Designs, Codes and Cryptography*, 20(1):41–64, 2000.

[114] Vahid Jafari and Mohammad Hossein Rezvani. Joint optimization of energy consumption and time delay in iot-fog-cloud computing environments using nsga-ii metaheuristic algorithm. *Journal of Ambient Intelligence and Humanized Computing*, 14(3):1675–1698, 2023.

[115] Anupriya Jain, Seema Sharma, and Sachin Sharma. Firefly algorithm. *Nature-Inspired Algorithms Applications*, pages 157–180, 2021.

[116] Saeed Ullah Jan, Fawad Qayum, and Habib Ullah Khan. Design and analysis of lightweight authentication protocol for securing iod. *Ieee access*, 9:69287–69306, 2021.

[117] Zohaib Jan, Farhad Ahamed, Wolfgang Mayer, Niki Patel, Georg Grossmann, Markus Stumptner, and Ana Kuusk. Artificial intelligence for industry 4.0: Systematic review of applications, challenges, and opportunities. *Expert Systems with Applications*, 216:119456, 2023.

[118] Saeed Javanmardi, Mohammad Shojafar, Reza Mohammadi, Valerio Persico, and Antonio Pescapè. S-fos: A secure workflow scheduling approach for performance optimization in sdn-based iot-fog networks. *Journal of Information Security and Applications*, 72:103404, 2023.

[119] Xiaoying Jia, Debiao He, Neeraj Kumar, and Kim-Kwang Raymond Choo. Authenticated key agreement scheme for fog-driven iot healthcare system. *Wireless Networks*, 25:4737–4750, 2019.

[120] Yu-Lin Jiang, Ya-Shu Chen, Su-Wei Yang, and Chia-Hsueh Wu. Energy-efficient task offloading for time-sensitive applications in fog computing. *IEEE Systems Journal*, 13(3):2930–2941, 2018.

[121] Stig Ole Johnsen, Trond Bakken, Aksel Andreas Transeth, Sture Holmstrøm, Mariann Merz, Esten Ingar Grøtli, Sigurd Robert Jacobsen, and Rune Storvold. Safety and security of drones in the oil and gas industry. In *e-proceedings of the 30th European Safety and Reliability Conference and 15th Probabilistic Safety Assessment and Management Conference (ESREL2020 PSAM15)*. Research Publishing Services, 2020.

[122] Aisha Kanwal Junejo, Nikos Komninos, and Julie A McCann. A secure integrated framework for fog-assisted internet-of-things systems. *IEEE Internet of Things Journal*, 8(8):6840–6852, 2020.

[123] Andreas Kamilaris and Francesc X Prenafeta-Boldú. Disaster monitoring using unmanned aerial vehicles and deep learning. *arXiv preprint arXiv:1807.11805*, 2018.

[124] Andreas Kamilaris and Francesc X Prenafeta-Boldú. Disaster monitoring using unmanned aerial vehicles and deep learning. *arXiv preprint arXiv:1807.11805*, 2018.

[125] Dervis Karaboga. An idea based on honey bee swarm for numerical optimization, technical report - tr06. *Technical Report, Erciyes University*, 01 2005.

[126] Abhishek Kashyap, Debasish Ghose, Prathyush P Menon, PB Sujit, and Kaushik Das. Uav aided dynamic routing of resources in a flood scenario. In *2019 International Conference on Unmanned Aircraft Systems (ICUAS)*, pages 328–335. IEEE, 2019.

[127] J. Kennedy and R. Eberhart. Particle swarm optimization. In *Proceedings of ICNN'95 - International Conference on Neural Networks*, volume 4, pages 1942–1948 vol.4, 1995.

[128] James Kennedy. Swarm intelligence. In *Handbook of nature-inspired and innovative computing: integrating classical models with emerging technologies*, pages 187–219. Springer, 2006.

[129] Abhishek Kesarwani and Pabitra Mohan Khilar. Development of trust based access control models using fuzzy logic in cloud computing. *Journal of King Saud University-Computer and Information Sciences*, 34(5):1958–1967, 2022.

[130] Niharika Keshari, Dinesh Singh, and Ashish Kumar Maurya. A survey on vehicular fog computing: Current state-of-the-art and future directions. *Vehicular Communications*, 38:100512, 2022.

[131] Maryam Keshavarznejad, Mohammad Hossein Rezvani, and Sepideh Adabi. Delay-aware optimization of energy consumption for task offloading in fog environments using metaheuristic algorithms. *Cluster Computing*, pages 1–29, 2021.

[132] Wasim Khan, Hannan Ansari, and Anwar Ahamed Shaikh. Log files utility for software maintenance. *International Journal of Advanced Research in Computer Engineering and Technology (IJARCET)*, 4, 2015.

[133] Wasim Khan and Mohammad Haroon. *A Pilot Study and Survey on Methods for Anomaly Detection in Online Social Networks*, pages 119–128. Springer, 2022.

[134] Wasim Khan and Mohammad Haroon. An unsupervised deep learning ensemble model for anomaly detection in static attributed social networks. *International Journal of Cognitive Computing in Engineering*, 3:153–160, 2022.

[135] Wasim Khan, Mohammad Haroon, Ahmad Neyaz Khan, Mohammad Kamrul Hasan, Asif Khan, Umi Asma Mokhtar, and Shayla Islam. Dvaegmm: Dual variational autoencoder with gaussian mixture model for anomaly detection on attributed networks. *IEEE Access*, 10:91160–91176, 2022.

[136] Wasim Khan, Mohammad Ishrat, Mohd Haleem, Ahmad Neyaz Khan, Mohammad Kamrul Hasan, and Nafees Akhter Farooqui. *An Extensive Study and Review on Dark Web Threats and Detection Techniques*, pages 202–219. IGI Global, 2023.

[137] Wasim Khana and Mohammad Haroonb. An exhaustive review on state-of-the-art techniques for anomaly detection on attributed networks. *Turk-*

ish Journal of Computer and Mathematics Education Vol, 12:6707–6722, 2021.

[138] Edris Khezri, Rebaz Othman Yahya, Hiwa Hassanzadeh, Mohsen Mohaidat, Sina Ahmadi, and Mohammad Trik. Dljsf: Data-locality aware job scheduling iot tasks in fog-cloud computing environments. *Results in Engineering*, page 101780, 2024.

[139] Sooyeon Kim, Jae Hyun Kwak, Byoungryul Oh, Da-Han Lee, and Duehee Lee. An optimal routing algorithm for unmanned aerial vehicles. *Sensors*, 21(4):1219, 2021.

[140] Neal Koblitz. Elliptic curve cryptosystems. *Mathematics of computation*, 48(177):203–209, 1987.

[141] Dongyoung Koo and Junbeom Hur. Privacy-preserving deduplication of encrypted data with dynamic ownership management in fog computing. *Future Generation Computer Systems*, 78:739–752, 2018.

[142] Harish Kundra, Wasim Khan, Meenakshi Malik, Kantilal Pitambar Rane, Rahul Neware, and Vishal Jain. Quantum-inspired firefly algorithm integrated with cuckoo search for optimal path planning. *International Journal of Modern Physics C*, 33:2250018, 2022.

[143] Mohammed Laroui, Boubakr Nour, Hassine Moungla, Moussa A Cherif, Hossam Afifi, and Mohsen Guizani. Edge and fog computing for iot: A survey on current research activities and future directions. *Computer Communications*, 180:210–231, 2021.

[144] Chanhyuk Lee, Jisoo Kim, Heedong Ko, and Byounghyun Yoo. Addressing iot storage constraints: A hybrid architecture for decentralized data storage and centralized management. *Internet of Things*, 25:101014, 2024.

[145] Dave Lewis, David Filip, and Harshvardhan J Pandit. An ontology for standardising trustworthy ai. *Factoring Ethics in Technology, Policy Making, Regulation and AI*, 65, 2021.

[146] Xiong Li, Jieyao Peng, Mohammad S Obaidat, Fan Wu, Muhammad Khurram Khan, and Chaoyang Chen. A secure three-factor user authentication protocol with forward secrecy for wireless medical sensor network systems. *IEEE Systems Journal*, 14(1):39–50, 2019.

[147] Shuang Liu, Jie Tian, Chao Zhai, and Tiantian Li. Joint computation offloading and resource allocation in vehicular edge computing networks. *Digital Communications and Networks*, 2022.

[148] Mariana Macedo, Hugo Siqueira, Elliackin Figueiredo, Clodomir Santana, Rodrigo C Lira, Anu Gokhale, and Carmelo Bastos-Filho. Overview

on binary optimization using swarm-inspired algorithms. *IEEE Access*, 9:149814–149858, 2021.

[149] Parikshit N Mahalle, Bayu Anggorojati, Neeli Rashmi Prasad, and Ramjee Prasad. Identity driven capability based access control (icac) scheme for the internet of things. In *2012 IEEE International Conference on Advanced Networks and Telecommunciations Systems (ANTS)*, pages 49–54. IEEE, 2012.

[150] Parikshit N Mahalle, Pravin A Thakre, Neeli Rashmi Prasad, and Ramjee Prasad. A fuzzy approach to trust based access control in internet of things. In *Wireless VITAE 2013*, pages 1–5. IEEE, 2013.

[151] Redowan Mahmud, Satish Narayana Srirama, Kotagiri Ramamohanarao, and Rajkumar Buyya. Quality of experience (qoe)-aware placement of applications in fog computing environments. *Journal of Parallel and Distributed Computing*, 132:190–203, 2019.

[152] Nasrin M Makbol, Bee Ee Khoo, and Taha H Rassem. Security analyses of false positive problem for the svd-based hybrid digital image watermarking techniques in the wavelet transform domain. *Multimedia Tools and Applications*, 77:26845–26879, 2018.

[153] Ismael Martinez, Abdelhakim Senhaji Hafid, and Abdallah Jarray. Design, resource management, and evaluation of fog computing systems: a survey. *IEEE Internet of Things Journal*, 8(4):2494–2516, 2020.

[154] Carles Martínez-García, Guillermo Navarro-Arribas, and Joan Borrell. Fuzzy role-based access control. *Information processing letters*, 111(10):483–487, 2011.

[155] Yassine Mekdad, Ahmet Aris, Leonardo Babun, Abdeslam El Fergougui, Mauro Conti, Riccardo Lazzeretti, and A. Selcuk Uluagac. A survey on security and privacy issues of uavs. *CoRR*, abs/2109.14442, 2021.

[156] Thomas S Messerges, Ezzat A Dabbish, and Robert H Sloan. Examining smart-card security under the threat of power analysis attacks. *IEEE transactions on computers*, 51(5):541–552, 2002.

[157] Adriana Mijuskovic, Alessandro Chiumento, Rob Bemthuis, Adina Aldea, and Paul Havinga. Resource management techniques for cloud/fog and edge computing: An evaluation framework and classification. *Sensors*, 21(5):1832, 2021.

[158] Victor S Miller. Use of elliptic curves in cryptography. In *Conference on the theory and application of cryptographic techniques*, pages 417–426. Springer, 1985.

[159] Sudip Misra and Niloy Saha. Detour: Dynamic task offloading in software-defined fog for iot applications. *IEEE Journal on Selected Areas in Communications*, 37(5):1159–1166, 2019.

[160] Eleftheria Mitka and Spiridon G Mouroutsos. Classification of drones. *Am. J. Eng. Res*, 6(7):36–41, 2017.

[161] Nader Mohamed, Jameela Al-Jaroodi, Sanja Lazarova-Molnar, and Imad Jawhar. Applications of integrated iot-fog-cloud systems to smart cities: A survey. *Electronics*, 10:2918, 2021.

[162] David R Monismith and Blayne E Mayfield. Slime mold as a model for numerical optimization. In *2008 IEEE swarm intelligence symposium*, pages 1–8. IEEE, 2008.

[163] Jose Moura and David Hutchison. Fog computing systems: State of the art, research issues and future trends, with a focus on resilience. *Journal of Network and Computer Applications*, 169:102784, 2020.

[164] Zahid Muhammad, Navrati Saxena, Ijaz Mansoor Qureshi, and Chang Wook Ahn. Hybrid artificial bee colony algorithm for an energy efficient internet of things based on wireless sensor network. *IETE Technical Review*, 34(sup1):39–51, 2017.

[165] SR Murugaiyan, D Chandramohan, T Vengattaraman, and P Dhavachelvan. A generic privacy breach preventing methodology for cloud based web service. *International Journal of Grid and High Performance Computing (IJGHPC)*, 6(3):53–84, 2014.

[166] Shiva Shankar Mutupuri, Vallabhaneni Preetam, Dudugu Aditya, Alapati Harishitha, Sruthi Sivarajan, and Satish Anamalamudi. Machine learning based malware detection for iot networks. In *2023 IEEE 15th International Conference on Computational Intelligence and Communication Networks (CICN)*, pages 298–303. IEEE, 2023.

[167] Ranesh Kumar Naha, Saurabh Garg, Dimitrios Georgakopoulos, Prem Prakash Jayaraman, Longxiang Gao, Yong Xiang, and Rajiv Ranjan. Fog computing: Survey of trends, architectures, requirements, and research directions. *IEEE access*, 6:47980–48009, 2018.

[168] Darius Nahavandi, Roohallah Alizadehsani, Abbas Khosravi, and U Rajendra Acharya. Application of artificial intelligence in wearable devices: Opportunities and challenges. *Computer Methods and Programs in Biomedicine*, 213:106541, 2022.

[169] Zhaojun Nan, Sheng Zhou, Yunjian Jia, and Zhisheng Niu. Joint task offloading and resource allocation for vehicular edge computing with result feedback delay. *IEEE Transactions on Wireless Communications*, 2023.

[170] Aadharsh Roshan Nandhakumar, Ayush Baranwal, Priyanshukumar Choudhary, Muhammed Golec, and Sukhpal Singh Gill. Edgeaisim: A toolkit for simulation and modelling of ai models in edge computing environments. *Measurement: Sensors*, 31:100939, 2024.

[171] BV Natesha and Ram Mohana Reddy Guddeti. Heuristic-based iot application modules placement in the fog-cloud computing environment. In *2018 IEEE/ACM international conference on utility and cloud computing companion (UCC Companion)*, pages 24–25. IEEE, 2018.

[172] Hira Nazir, Imran Sarwar Bajwa, Muhammad Samiullah, Waheed Anwar, and Muhammad Moosa. Robust secure color image watermarking using 4d hyperchaotic system, dwt, hbd, and svd based on improved foa algorithm. *Security and Communication Networks*, 2021:1–17, 2021.

[173] Nadia Nedjah and Luneque Silva Junior. Review of methodologies and tasks in swarm robotics towards standardization. *Swarm and Evolutionary Computation*, 50:100565, 2019.

[174] Jianbing Ni, Kuan Zhang, Xiaodong Lin, and Xuemin Shen. Securing fog computing for internet of things applications: Challenges and solutions. *IEEE Communications Surveys and Tutorials*, 20(1):601–628, 2017.

[175] Lina Ni, Jinquan Zhang, Changjun Jiang, Chungang Yan, and Kan Yu. Resource allocation strategy in fog computing based on priced timed petri nets. *ieee internet of things journal*, 4(5):1216–1228, 2017.

[176] Nikos Th Nikolinakos. *Ethical Principles for Trustworthy AI*, pages 101–166. Springer, 2023.

[177] Zhaolong Ning, Jun Huang, and Xiaojie Wang. Vehicular fog computing: Enabling real-time traffic management for smart cities. *IEEE Wireless Communications*, 26(1):87–93, 2019.

[178] Michele Nitti, Roberto Girau, and Luigi Atzori. Trustworthiness management in the social internet of things. *IEEE Transactions on knowledge and data engineering*, 26(5):1253–1266, 2013.

[179] AM Odlyzko. Discrete logarithms and their cryptographic significance. advances in cryptology: Proceedings of eurocrypt 84, 1985.

[180] Andrew Odlyzko. Discrete logarithms: The past and the future. *Towards a Quarter-Century of Public Key Cryptography: A Special Issue of DESIGNS, CODES AND CRYPTOGRAPHY An International Journal. Volume 19, No. 2/3 (2000)*, pages 59–75, 2000.

[181] Sunday Oyinlola Ogundoyin and Ismaila Adeniyi Kamil. Optimization techniques and applications in fog computing: An exhaustive survey. *Swarm and Evolutionary Computation*, 66:100937, 2021.

[182] World Health Organization. Ethics and governance of artificial intelligence for health: Who guidance. 2021.

[183] Aafaf Ouaddah, Hajar Mousannif, Anas Abou Elkalam, and Abdellah Ait Ouahman. Access control in the internet of things: Big challenges and new opportunities. *Computer Networks*, 112:237–262, 2017.

[184] Omar Sami Oubbati, Mohammed Atiquzzaman, Pascal Lorenz, Abdullah Baz, and Hosam Alhakami. Search: An sdn-enabled approach for vehicle path-planning. *IEEE Transactions on Vehicular Technology*, 69(12):14523–14536, 2020.

[185] Omar Sami Oubbati, Mohammad Mozaffari, Noureddine Chaib, Pascal Lorenz, Mohammed Atiquzzaman, and Abbas Jamalipour. Ecad: Energy-efficient routing in flying ad hoc networks. *International Journal of Communication Systems*, 32(18):e4156, 2019.

[186] Mrutyunjaya Panda. *Multi-objective Swarm Intelligence Theoretical Advances and Applications*. 01 2015.

[187] Kevin M Passino. Biomimicry of bacterial foraging for distributed optimization and control. *IEEE control systems magazine*, 22(3):52–67, 2002.

[188] K.M. Passino. Biomimicry of bacterial foraging for distributed optimization and control. *IEEE Control Systems Magazine*, 22(3):52–67, 2002.

[189] RK Pateriya and Shreeja Vasudevan. Elliptic curve cryptography in constrained environments: a review. In *2011 International Conference on Communication Systems and Network Technologies*, pages 120–124. IEEE, 2011.

[190] Anand Paul, Hameed Pinjari, Won-Hwa Hong, Hyun Cheol Seo, and Seungmin Rho. Fog computing-based iot for health monitoring system. *Journal of Sensors*, 2018:1386470, 2018.

[191] NR Paul and D Paul Raj. Enhanced trust based access control for multi-cloud environment. *Computers, Materials & Continua*, 69(3), 2021.

[192] AAN Perwira Redi, Nandini N Sekaringtyas, Handina B Astiana, Rahmat Inca Liperda, Anna Maria Sri Asih, and Bertha Maya Sopha. Two echelon vehicle routing problem for disaster mapping process using co-operated vehicles assignment. In *Smart and Sustainable Supply Chain and Logistics—Challenges, Methods and Best Practices: Volume 2*, pages 233–242. Springer, 2023.

[193] C S R Prabhu. *Fog computing, deep learning and big data analytics-research directions*. Springer, 2019.

[194] M Praveen Kumar Reddy and M Rajasekhara Babu. Energy efficient cluster head selection for internet of things. *New Review of Information Networking*, 22(1):54–70, 2017.

[195] Guntur Dharma Putra, Volkan Dedeoglu, Salil S Kanhere, and Raja Jurdak. Trust management in decentralized iot access control system. In *2020 IEEE international conference on blockchain and cryptocurrency (ICBC)*, pages 1–9. IEEE, 2020.

[196] Gaoyuan Quan, Zhongyuan Yao, Longfei Chen, Yonghao Fang, Weihua Zhu, Xueming Si, and Min Li. A trusted medical data sharing framework for edge computing leveraging blockchain and outsourced computation. *Heliyon*, 9(12), 2023.

[197] Anichur Rahman, Md Jahidul Islam, Shahab S Band, Ghulam Muhammad, Kamrul Hasan, and Prayag Tiwari. Towards a blockchain-sdn-based secure architecture for cloud computing in smart industrial iot. *Digital Communications and Networks*, 9(2):411–421, 2023.

[198] S Suman Rajest, Bhopendra Singh, Ahmed J Obaid, R Regin, and Karthikeyan Chinnusamy. *Advances in Artificial and Human Intelligence in the Modern Era*. IGI Global, 2023.

[199] Diksha Rangwani and Hari Om. A secure user authentication protocol based on ecc for cloud computing environment. *Arabian Journal for Science and Engineering*, 46:3865–3888, 2021.

[200] Diksha Rangwani, Dipanwita Sadhukhan, Sangram Ray, Muhammad Khurram Khan, and Mou Dasgupta. An improved privacy preserving remote user authentication scheme for agricultural wireless sensor network. *Transactions on Emerging Telecommunications Technologies*, 32(3):e4218, 2021.

[201] Shalli Rani, Gautam Srivastava, et al. Secure hierarchical fog computing-based architecture for industry 5.0 using an attribute-based encryption scheme. *Expert Systems with Applications*, 235:121180, 2024.

[202] Sudesh Rani and Poonam Saini. Fog computing: applications and secure data aggregation. *Handbook of Computer Networks and Cyber Security: Principles and Paradigms*, pages 475–492, 2020.

[203] Al-Huthaifi Rasha, Tianrui Li, Wei Huang, Jin Gu, and Chongshou Li. Federated learning in smart cities: Privacy and security survey. *Information Sciences*, 2023.

[204] Motashim Rasool and Wasim Khan. Big data: Study in structured and unstructured data. *HCTL Open International Journal of Technology Innovations and Research (IJTIR)*, 14:1–6, 2015.

[205] Salman Raza, Shangguang Wang, Manzoor Ahmed, Muhammad Rizwan Anwar, Muhammad Ayzed Mirza, and Wali Ullah Khan. Task offloading and resource allocation for iov using 5g nr-v2x communication. *IEEE Internet of Things Journal*, 9(13):10397–10410, 2021.

[206] Abdul Razaque, Yaser Jararweh, Bandar Alotaibi, Munif Alotaibi, Salim Hariri, and Muder Almiani. Energy-efficient and secure mobile fog-based cloud for the internet of things. *Future Generation Computer Systems*, 127:1–13, 2022.

[207] V Dinesh Reddy and GSVRK Rao. Meta-heuristic approaches to solve shortest lattice vector problem. *Journal of Discrete Mathematical Sciences and Cryptography*, 24(1):81–91, 2021.

[208] Agoston Restas et al. Drone applications for supporting disaster management. *World Journal of Engineering and Technology*, 3(03):316, 2015.

[209] Agoston Restas et al. Drone applications for supporting disaster management. *World Journal of Engineering and Technology*, 3(03):316, 2015.

[210] Zahra Rezazadeh, Dadmehr Rahbari, and Mohsen Nickray. Optimized module placement in iot applications based on fog computing. In *Electrical engineering (ICEE), Iranian conference on*, pages 1553–1558. IEEE, 2018.

[211] Samad Rostampour, Masoumeh Safkhani, Ygal Bendavid, and Nasour Bagheri. Eccbap: A secure ecc-based authentication protocol for iot edge devices. *Pervasive and Mobile Computing*, 67:101194, 2020.

[212] Karina A Roundtree, Michael A Goodrich, and Julie A Adams. Transparency: Transitioning from human–machine systems to human-swarm systems. *Journal of Cognitive Engineering and Decision Making*, 13:171–195, 2019.

[213] Indranil Roy, Reshmi Mitra, Nick Rahimi, and Bidyut Gupta. Efficient non-dht-based rc-based architecture for fog computing in healthcare 4.0. *IoT*, 4:131–149, 2023.

[214] Dipanwita Sadhukhan, Sangram Ray, GP Biswas, Muhammad Khurram Khan, and Mou Dasgupta. A lightweight remote user authentication scheme for iot communication using elliptic curve cryptography. *The Journal of Supercomputing*, 77:1114–1151, 2021.

[215] Yosra Ben Saied, Alexis Olivereau, Djamal Zeghlache, and Maryline Laurent. Trust management system design for the internet of things: A context-aware and multi-service approach. *Computers & Security*, 39:351–365, 2013.

[216] Farah Ait Salaht, Frédéric Desprez, and Adrien Lebre. An overview of service placement problem in fog and edge computing. *ACM Computing Surveys (CSUR)*, 53(3):1–35, 2020.

[217] Sasa Sambolek and Marina Ivasic-Kos. Person detection in drone imagery. In *2020 5th International Conference on Smart and Sustainable Technologies (SpliTech)*, pages 1–6. IEEE, 2020.

[218] Arup Kumar Sarker, Md Khairul Islam, Yuan Tian, and Geoffrey Fox. Mvam: Multi-variant attacks on memory for iot trust computing. In *Proceedings of Cyber-Physical Systems and Internet of Things Week 2023*, pages 13–18. 2023.

[219] Melanie Schranz, Gianni A Di Caro, Thomas Schmickl, Wilfried Elmenreich, Farshad Arvin, Ahmet Şekercioğlu, and Micha Sende. Swarm intelligence and cyber-physical systems: concepts, challenges and future trends. *Swarm and Evolutionary Computation*, 60:100762, 2021.

[220] Reva Schwartz, Apostol Vassilev, Kristen Greene, Lori Perine, Andrew Burt, and Patrick Hall. Towards a standard for identifying and managing bias in artificial intelligence. *NIST special publication*, 1270, 2022.

[221] Aritro Sengupta, Amit Singh, Pankaj Kumar, and Tapobrata Dhar. A secure and improved two factor authentication scheme using elliptic curve and bilinear pairing for cyber physical systems. *Multimedia Tools and Applications*, 81(16):22425–22448, 2022.

[222] Ali Shakarami, Hamid Shakarami, Mostafa Ghobaei-Arani, Elaheh Nikougoftar, and Mohammad Faraji-Mehmandar. Resource provisioning in edge/fog computing: A comprehensive and systematic review. *Journal of Systems Architecture*, 122:102362, 2022.

[223] Hazim Shakhatreh, Ahmad H Sawalmeh, Ala Al-Fuqaha, Zuochao Dou, Eyad Almaita, Issa Khalil, Noor Shamsiah Othman, Abdallah Khreishah, and Mohsen Guizani. Unmanned aerial vehicles (uavs): A survey on civil applications and key research challenges. *Ieee Access*, 7:48572–48634, 2019.

[224] Hazim Shakhatreh, Ahmad H Sawalmeh, Ala Al-Fuqaha, Zuochao Dou, Eyad Almaita, Issa Khalil, Noor Shamsiah Othman, Abdallah Khreishah, and Mohsen Guizani. Unmanned aerial vehicles (uavs): A survey on civil

applications and key research challenges. *Ieee Access*, 7:48572–48634, 2019.

[225] Liping Shi, Néstor J Hernández Marcano, and Rune Hylsberg Jacobsen. A review on communication protocols for autonomous unmanned aerial vehicles for inspection application. *Microprocessors and Microsystems*, 86:104340, 2021.

[226] Joseph H Silverman. The xedni calculus and the elliptic curve discrete logarithm problem. *Designs, Codes and Cryptography*, 20(1):5–40, 2000.

[227] Inderpal Singh and Balraj Singh. Access management of iot devices using access control mechanism and decentralized authentication: A review. *Measurement: Sensors*, 25:100591, 2023.

[228] Laxmanika Singh, Amit Kumar Singh, and Pradeep Kumar Singh. Secure data hiding techniques: a survey. *Multimedia Tools and Applications*, 79:15901–15921, 2020.

[229] Nathalie A Smuha. The eu approach to ethics guidelines for trustworthy artificial intelligence. *Computer Law Review International*, 20:97–106, 2019.

[230] Jonathan Sorenson. An introduction to prime number sieves. university of wisconsin-madison. *Computer Sciences Department*, 1990.

[231] Jonathan P Sorenson. The pseudosquares prime sieve. In *International Algorithmic Number Theory Symposium*, pages 193–207. Springer, 2006.

[232] K Sowjanya, Mou Dasgupta, Sangram Ray, and Mohammad S Obaidat. An efficient elliptic curve cryptography-based without pairing kpabe for internet of things. *IEEE Systems Journal*, 14(2):2154–2163, 2019.

[233] R Leena Sri and V Divya. *Intelligent energy-aware decision-making at the edge in healthcare using fog infrastructure*, pages 87–107. Elsevier, 2020.

[234] Bernd Carsten Stahl. *Artificial intelligence for a better future: an ecosystem perspective on the ethics of AI and emerging digital technologies*. Springer Nature, 2021.

[235] Nicolas-Alin Stoian. Machine learning for anomaly detection in iot networks: Malware analysis on the iot-23 data set. B.S. thesis, University of Twente, 2020.

[236] Qingtang Su, Decheng Liu, Zihan Yuan, Gang Wang, Xiaofeng Zhang, Beijing Chen, and Tao Yao. New rapid and robust color image watermarking technique in spatial domain. *IEEE Access*, 7:30398–30409, 2019.

[237] Weifeng Sun, Min Tang, Lijun Zhang, Zhiqiang Huo, and Lei Shu. A survey of using swarm intelligence algorithms in iot. *Sensors*, 20:1420, 03 2020.

[238] Farheen Syed, Sachin Kumar Gupta, Saeed Hamood Alsamhi, Mamoon Rashid, and Xuan Liu. A survey on recent optimal techniques for securing unmanned aerial vehicles applications. *Transactions on Emerging Telecommunications Technologies*, 32(7):e4133, 2021.

[239] Nazia Tabassum and CRK Reddyy. Review on qos and security challenges associated with the internet of vehicles in cloud computing. *Measurement: Sensors*, 27:100562, 2023.

[240] Chaogang Tang, Shixiong Xia, Qing Li, Wei Chen, and Weidong Fang. Resource pooling in vehicular fog computing. *Journal of Cloud Computing*, 10:1–14, 2021.

[241] Tullio Joseph Tanzi, Madhu Chandra, Jean Isnard, Daniel Camara, Olivier Sebastien, and Fanilo Harivelo. Towards" drone-borne" disaster management: future application scenarios. In *XXIII ISPRS Congress, Commission VIII (Volume III-8)*, volume 3, pages 181–189. Copernicus GmbH, 2016.

[242] Tribhuwan Kumar Tewari and Vikas Saxena. An improved and robust dct based digital image watermarking scheme. *International Journal of Computer Applications*, 3(1):28–32, 2010.

[243] Rohit Thanki, Ashish Kothari, and Surekha Borra. Hybrid, blind and robust image watermarking: Rdwt–nsct based secure approach for telemedicine applications. *Multimedia Tools and Applications*, 80(18):27593–27613, 2021.

[244] Usha Vadde and Vijaya Sri Kompalli. Energy efficient service placement in fog computing. *PeerJ Computer Science*, 8:e1035, 2022.

[245] Huaqun Wang, Zhiwei Wang, and Josep Domingo-Ferrer. Anonymous and secure aggregation scheme in fog-based public cloud computing. *Future Generation Computer Systems*, 78:712–719, 2018.

[246] Xiaojie Wang, Zhaolong Ning, and Lei Wang. Offloading in internet of vehicles: A fog-enabled real-time traffic management system. *IEEE Transactions on Industrial Informatics*, 14(10):4568–4578, 2018.

[247] Mohammad Wazid, Ashok Kumar Das, Neeraj Kumar, and Athanasios V Vasilakos. Design of secure key management and user authentication scheme for fog computing services. *Future Generation Computer Systems*, 91:475–492, 2019.

[248] Tsu-Yang Wu, Zhiyuan Lee, Lei Yang, Jia-Ning Luo, and Raylin Tso. Provably secure authentication key exchange scheme using fog nodes in vehicular ad hoc networks. *The Journal of Supercomputing*, 77:6992–7020, 2021.

[249] Tsu-Yang Wu, Tao Wang, Yu-Qi Lee, Weimin Zheng, Saru Kumari, and Sachin Kumar. Improved authenticated key agreement scheme for fog-driven iot healthcare system. *Security and Communication Networks*, 2021:1–16, 2021.

[250] Jianwen Xu, Kaoru Ota, and Mianxiong Dong. Luna: Lightweight uav navigation based on airborne vision for disaster management. In *2019 International conference on Internet of Things (iThings) and IEEE green computing and communications (GreenCom) and IEEE cyber, physical and social computing (CPSCom) and IEEE Smart Data (SmartData)*, pages 315–322. IEEE, 2019.

[251] Wanli Xue, Yiran Shen, Chengwen Luo, Weitao Xu, Wen Hu, and Aruna Seneviratne. A differential privacy-based classification system for edge computing in iot. *Computer Communications*, 182:117–128, 2022.

[252] Jen-Ho Yang and Chin-Chen Chang. An id-based remote mutual authentication with key agreement scheme for mobile devices on elliptic curve cryptosystem. *Computers and security*, 28(3-4):138–143, 2009.

[253] Yang Yang, Xiliang Luo, Xiaoli Chu, and Ming-Tuo Zhou. *Fog-enabled intelligent IoT systems*. Springer, 2020.

[254] Zhen Yang, Qingwei Sun, Yunliang Qi, Shouliang Li, and Fengyuan Ren. A hyper-chaotically encrypted robust digital image watermarking method with large capacity using compress sensing on a hybrid domain. *Entropy*, 24(10):1486, 2022.

[255] Ashkan Yousefpour, Caleb Fung, Tam Nguyen, Krishna Kadiyala, Fatemeh Jalali, Amirreza Niakanlahiji, Jian Kong, and Jason P Jue. All one needs to know about fog computing and related edge computing paradigms: A complete survey. *Journal of Systems Architecture*, 98:289–330, 2019.

[256] Aditi Zear and Pradeep Kumar Singh. Secure and robust color image dual watermarking based on lwt-dct-svd. *Multimedia Tools and Applications*, 81(19):26721–26738, 2022.

[257] Ouarda Zedadra, Antonio Guerrieri, Nicolas Jouandeau, Giandomenico Spezzano, Hamid Seridi, and Giancarlo Fortino. Swarm intelligence-based algorithms within iot-based systems: A review. *J. Parallel Distributed Comput.*, 122:173–187, 2018.

[258] Zhenyu Zhou, Haijun Liao, Xiaoyan Wang, Shahid Mumtaz, and Jonathan Rodriguez. When vehicular fog computing meets autonomous driving: Computational resource management and task offloading. *IEEE Network*, 34(6):70–76, 2020.

[259] Chao Zhu, Jin Tao, Giancarlo Pastor, Yu Xiao, Yusheng Ji, Quan Zhou, Yong Li, and Antti Ylä-Jääski. Folo: Latency and quality optimized task allocation in vehicular fog computing. *IEEE Internet of Things Journal*, 6(3):4150–4161, 2018.

[260] Ting Zhu, Wen Qu, and Wenliang Cao. An optimized image watermarking algorithm based on svd and iwt. *The Journal of Supercomputing*, 78(1):222–237, 2022.

[261] Xiangyu Zhuo, Tobias Koch, Franz Kurz, Friedrich Fraundorfer, and Peter Reinartz. Automatic uav image geo-registration by matching uav images to georeferenced image data. *Remote Sensing*, 9(4):376, 2017.

[262] Yulong Zou, Jia Zhu, Tonghua Wu, Haiyan Guo, and Hao Wei. Cooperative drone communications for space-air-ground integrated networks. *IEEE Network*, 35(5):100–106, 2021.

Index

S

T

U

For Product Safety Concerns and Information please contact our EU representative GPSR@taylorandfrancis.com Taylor & Francis Verlag GmbH, Kaufingerstraße 24, 80331 München, Germany